Plumbing Services Series

Gasfitting

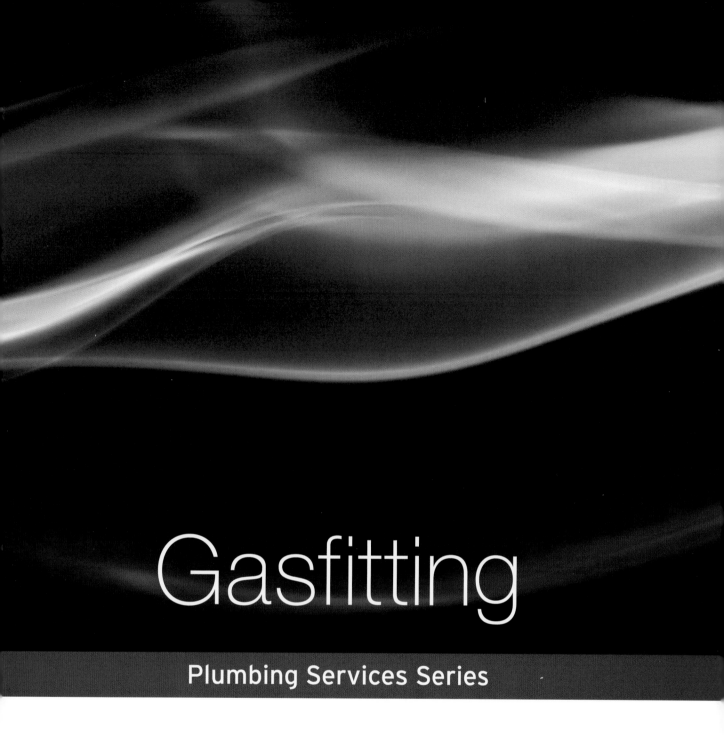

Gasfitting

Plumbing Services Series

2nd Edition

Michael Burn, Peter Miles

R.J. Puffett, L.J. Hossack, J. Stone

Reprinted 2019, 2020, 2021, 2022, 2023

National Library of Australia Cataloguing-in-Publication Data

Title: Gasfitting / Michael Burn ... [et al.]
Edition: 2nd ed.
ISBN: 9780070278912 (pbk.)
Series: Plumbing services.
Notes: Includes index.
Subjects: Gas-fitting—Handbooks, manuals, etc.
Other Authors/Contributors:
             Burn, Michael Colin.

Dewey Number: 696.2

Published in Australia by
McGraw-Hill Australia Pty Ltd
Level 33, 680 George Street, Sydney NSW 2000
Sponsoring Editor: Michael Buhagiar
Managing Editor: Kathryn Fairfax
Development Editor: Amanda Evans
Production Editors: Eleanna Raissis and Claire Linsdell
Editor: Nichole McKenzie
Permissions Editor: Haidi Bernhardt
Editorial Coordinator: Vanessa Cork
Illustrator: Aptara Inc., India
Proofreaders: Ron Gillett and Terence Townsend
Indexer: Shelley Barons
Art Director: Astred Hicks
Cover design: Luke Causby, Blue Cork
Internal design: Natalie Bowra and David Rosemeyer
Typeset in Baskerville 10.5/12 by Midland Typesetters, Australia
Printed in Singapore on 90 gsm matt art Markono Print Media Pte Ltd

Contents

Preface

The second edition of *Gasfitting* brings this classic text up to date for the requirements of the modern teaching environment. Written in a clear, accessible style, and with an abundance of illustrations and photographs, it covers all the essential competencies of the gasfitting component of the Plumbing and Services Training Package, and provides a depth of underlying knowledge to help the student towards a thorough understanding of the subject.

The material presented here is able to be applied generally. However, it should be noted that existing local conditions and requirements may differ. Where such a situation occurs, the local regulations must take precedence over the information supplied here. Climatic conditions, for example, vary considerably and may dictate local variations in regulations and practice. Differences between the states and between countries must also be taken into account. The principles, however, remain the same, and the text can therefore be read in conjunction with these requirements.

Although the gas industry in Australia is controlled by companies that vary from large government and public utilities to small councils and private companies, a national standard for the installation and testing of gas appliances issued by the Australian Gas Association is generally followed by all gas companies. A copy of the current code for the installation of gas burning appliances and equipment is considered essential.

The basic hand skills required to carry out an installation and fix gas appliances are very similar to those required to carry out a plumbing installation. Additional knowledge and skills must be acquired by the plumber in order to be capable of installing, commissioning and servicing gas appliances to the standards demanded in the gas industry. The information contained within this volume is aimed, in general, at the basic trade level and therefore is suitable for students seeking registration as gasfitters. The text will also assist the licensed gasfitter to understand and correctly interpret the standards laid down in the current code for the installation of gas burning appliances.

We would like to thank the teachers who generously gave their expert advice on this project, and those who assisted with the editing and production of the manuscript.

Acknowledgements

Special thanks are due to Gary Borg (Granville TAFE) for his helpful advice at an early stage of the project; Andrew Craine (Gymea TAFE) for his invaluable technical suggestions; and David Harrison and Tim Comerford (Meadowbank TAFE) for their generosity with their time and advice with respect to the photos.

The publisher would like to thank the following for permission to reproduce their images:
- Electrolux Home Products Pty Ltd, for Westinghouse products (figs 11.1, 11.3 and 11.4)
- Arisit Pty Limited (Australia) for the Ariston range cooker (fig. 11.2)
- Fisher & Paykel Appliances Limited for the gas hotplates (fig. 11.5)
- Aber Holdings Limited for the GEO wall furnace (fig. 11.21).

E-student/E-instructor

Online *LearningCentre*

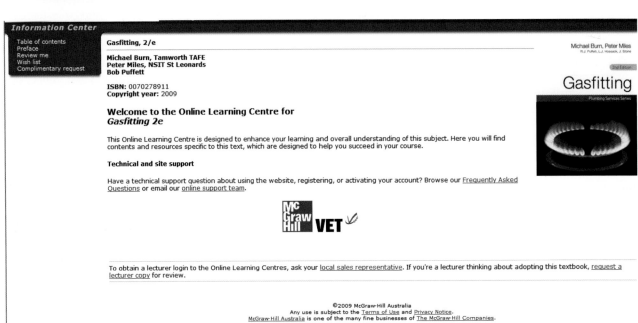

Information Center

Table of contents
Preface
Review me
Wish list
Complimentary request

Gasfitting, 2/e

Michael Burn, Tamworth TAFE
Peter Miles, NSIT St Leonards
Bob Puffett

ISBN: 0070278911
Copyright year: 2009

Welcome to the Online Learning Centre for
Gasfitting 2e

This Online Learning Centre is designed to enhance your learning and overall understanding of this subject. Here you will find contents and resources specific to this text, which are designed to help you succeed in your course.

Technical and site support

Have a technical support question about using the website, registering, or activating your account? Browse our Frequently Asked Questions or email our online support team.

To obtain a lecturer login to the Online Learning Centres, ask your local sales representative. If you're a lecturer thinking about adopting this textbook, request a lecturer copy for review.

Michael Burn, Peter Miles
R.J. Puffett, L.J. Hossack, J. Stone

2nd Edition

Gasfitting

Plumbing Services Series

The Online Learning Centre (OLC) that accompanies this text is an integrated online product to assist you in getting the most from your course, providing a powerful learning experience beyond the printed page. Each component of the OLC can be found in both the student and instructor editions.

PowerPoint® slides

A set of PowerPoint® slides accompanies each chapter and features items that provide a lecture outline, plus key figures and tables from the text.

Art Library

All the illustrations from the text in a convenient ready-to-use format.

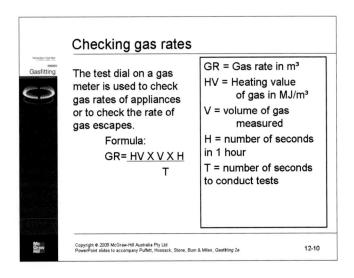

Checking gas rates

The test dial on a gas meter is used to check gas rates of appliances or to check the rate of gas escapes.
  Formula:
  $$GR = \frac{HV \times V \times H}{T}$$

GR = Gas rate in m³
HV = Heating value of gas in MJ/m³
V = volume of gas measured
H = number of seconds in 1 hour
T = number of seconds to conduct tests

Copyright © 2009 McGraw-Hill Australia Pty Ltd
PowerPoint slides to accompany Puffett, Hossack, Stone, Burn & Miles, *Gasfitting 2e*

12-10

CHAPTER ONE

Types of gases marketed in Australia

ORIGIN AND SOURCE OF SUPPLY OF GASES

There are many different types of gases that have been and that are currently being marketed throughout Australia. It is essential that the gasfitter understands the characteristics of the various gases and that the appliances are installed and commissioned for the correct gas type.

The type of gas can vary within the reticulated system; some of the larger companies may supply several different types of gases. The gas type should be determined before the purchase of appliances.

The following types of gases are currently marketed in Australia:

1   natural gas (NG)

2   liquefied petroleum gas (LPG)

3   tempered liquefied petroleum gas (TLP)

4   simulated natural gas (SNG)

Towns gas was marketed in Australia until the 1970s and processed natural gas was available until the 1990s.

Natural gas (NG)

It is generally believed that natural gas was formed from decaying organic matter. Over a period of up to 200 million years, this organic matter has been compressed by layer after layer of sediment built up over the centuries. The build-up of pressure and heat on the organic matter caused a chemical action which changed it into gas and oil. Although the sediment built up in layers, earthquakes have caused these layers to buckle and rise to form mountains and valleys. Natural gas became trapped in dome-like structures several thousand metres below the ground (Fig 1.1).

Natural gas was first found in 1900 by accident at Roma in Queensland during drilling for underground water. The gas was used to light the town's street lamps for 10 days before the supply failed. In 1961, natural gas was connected to the Roma Power Station. This was Australia's first commercial gas project.

Natural gas was introduced in Adelaide, Brisbane and Melbourne in 1969. Sydney gained natural gas in 1976 after the completion of the 1160 km pipeline from the Moomba gas fields in South Australia.

Geologists carry out surveys to find areas of the continental shelf and the inland sedimentary basins that may contain natural gas or oil. Seismographic recorders are used to locate the dome-like structures which may contain gas or oil.

A drilling rig is then used to drill an exploratory well to determine whether or not a commercial volume of gas or oil exists. Productive wells are linked together in a gathering network and then piped to a treatment plant.

The treatment plant separates the free liquids, reduces the CO_2 content to an acceptable level and removes the heavy hydrocarbons.

From the 1970s to 2000, natural gas increased as a proportion of Australia's total primary energy from 2% to 20% with the majority of it being used by industry. North West Shelf gas from Western Australia is now exported as liquid natural gas (LNG).

FIG 1.1 Location of natural gas wells

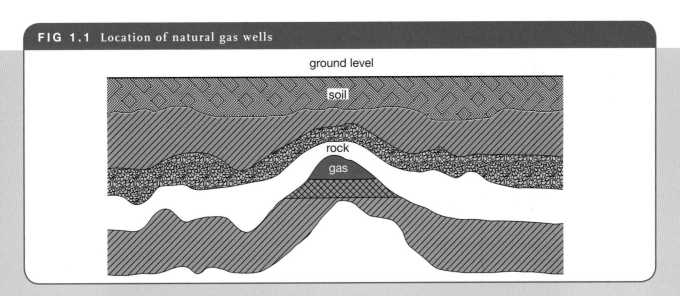

Liquefied petroleum gas (LPG)

Liquefied petroleum gas is the name given to a group of gases—propane is the one which is normally marketed in Australia. Propane and butane are by-products of natural gas wells or oil refineries. Propane can be converted to a liquid with the application of moderate pressures, and because of its low boiling point it will vaporise in the cylinder to replace the gas which is being used. Propane can be stored and transported in steel pressure vessels. When stored as a liquid, propane is a highly concentrated source of energy, making it an ideal fuel for use in areas where reticulated gases are not available.

A propane-butane mix is popular as an alternative fuel for vehicles, known as auto gas because it is considerably cheaper than petrol.

Non-odorised LPG mixes are now used as a propellant in aerosols to replace the chlorofluorocarbons (CFCs).

Tempered liquefied petroleum gas (TLP)

This gas was developed to replace towns gas in small provincial cities. A small and inexpensive plant can produce a substitute for towns gas by using propane or butane as a feedstock and mixing it with precise amounts of air until it has characteristics similar to towns gas.

Simulated natural gas (SNG)

Simulated natural gas is produced by a plant very similar to that used for producing TLP. Propane or butane is used as a feedstock which is mixed with air until it has the essential characteristics of natural gas. SNG is used as a holding gas in areas where it is anticipated that natural gas will become available. When natural gas is available, the installations and appliances can be connected with only a few minor adjustments being necessary.

Towns gas (TG)

From the start of the Australian gas industry in Sydney in 1841, towns gas was manufactured by carbonising coal. The carbonising of coal took place in retorts where it was heated to a very high temperature for a long period of time. This type of towns gas had to go through a very lengthy refining process before it was ready for distribution to the customer.

The gas was distributed in a wet state and so a means of removing the condensate from the installation had to be provided (i.e. tail pipes and siphon pits).

Towns gas manufactured from coal began to be phased out in the early 1960s with the introduction of catalytic reforming plants using oil as a feedstock. These catalytic reforming plants produce a gas similar to that from coal gas plants, but require far less manpower or time to manufacture a large volume of gas. Both these gases are toxic.

Processed natural gas (PNG)

The Australian Gas Light Company (AGL) of Sydney was using natural gas as a feedstock to produce a simulated towns gas known as processed natural gas. This provided an interim gas supply to consumers so that they could continue to use their towns gas appliances until they were converted to natural gas.

TABLE 1.1 Properties of gases (%)	NG	TLP	SNG	Butane	Propane
Hydrogen	–	–	–	–	–
Oxygen	0.001	14.7	9.5	–	–
Nitrogen	0.32	59	38.04	–	–
Carbon monoxide	–	–	–	–	–
Carbon dioxide	2.57	–	–	–	–
Methane	90.9	–	–	–	–
Propane	1.02	25.6	51.24	–	97.6
Butane	0.18	0.2	0.41	–	2.4
Other hydrocarbons	5.009	0.5	2.81	–	–

TABLE 1.2 Characteristics of gases	NG	TLP	SNG	Butane	Propane
Heating value MJ^3	38	25	48.5	120	96
Relative density	0.6	1.1	1.2	2	1.56
Burning velocity	400 mm/s	–	–	500mm/s	500mm/s
Air requirements	10:1 gas	6:1 gas	10:1 gas	–	24:1 gas
Ignition temperature	680°C	–	–	480°C	490°C
Flammability limits	5 to 14%	2.5 to 9.5%	2.5 to 9.5%	1.5–8.5%	2.5 to 9.5%
Toxicity	No	No	No	No	No
Odour added	Yes	Yes	Yes	Yes	Yes

PROPERTIES AND CHARACTERISTICS OF GASES

Table 1.1 details the constituents of different types of gases. These constituents vary from one production plant to another. The plumber and gasfitter should be aware of the main constituents and any particular effect they have on the characteristics of the gas, which are shown in Table 1.2.

Heating value (*HV*)

Heating value is the measurement of the amount of energy liberated when one cubic metre of gas, as defined, is completely burnt in air under standard test conditions. The temperature and pressure of the air for combustion and the resultant products of combustion are at 15°C and 101.325 kPa, respectively. Although the gas used by the customer is measured by volume, they are charged for the amount of energy used. Therefore the *HV* is very accurately monitored. The *HV* is measured by the number of megajoules (MJ) contained in a cubic metre of gas. For example the *HV* of natural gas is 38 MJ/mm³.

Relative density (*RD*)

Relative density is the weight of gas relative to the weight of air. It affects two important factors:

Safety

All gases will diffuse in air. Gases that are lighter than air rise, mix with air and dissipate quite quickly. Gases that are considerably heavier than air, such as propane, may drop to the lowest point. If propane builds up, on the ground or in a confined area, dissipation will be slower, as air may be in contact only with the surface of the gas and a dangerous situation may be created.

Working pressure

When gas is pushed out of the injector into the burner, it has work to do; to do this work effectively, heavier gases such as propane use a higher working pressure.

Burning velocity (flame propagation)

This is the speed at which the flame can consume the gas and is usually measured in millimetres per second (mm/s). If natural gas was placed in a glass tube 400 mm long and then lit at one end, it would take one second for the flame to travel to the other end and burn the gas in the tube. This is slow in comparison with towns gas which could travel 1050 mm in one second. With slow burning gases such as natural gas, there is a tendency for the gas to 'lift off' at the burner unless some means of flame retention is used (see Chapter 3, Figs 3.10 to 3.13).

Air requirements

Air must be available for combustion to take place. The ratio of air to gas illustrated in Table 1.2 will ensure complete combustion takes place.

Ignition temperatures

For the combustion process to take place, the air–gas mixture must of course be ignited. The ignition temperature varies depending upon the type of gas. As can be seen from Table 1.2, natural gas requires a higher temperature to ignite than towns gas or propane. Because of this, modern appliances generally use a high voltage spark or a pilot flame for ignition. The actual ignition temperature of the gas varies slightly depending upon the air–gas mixture.

Flammability limits (explosive limits)

Not all air–gas mixtures are combustible so it can be said, therefore, that air–gas mixtures are only combustible or explosive within certain limits.

Lower explosive limit (LEL)

The lower explosive limit is the minimum percentage of gas mixed with air that will burn or explode; for example, the LEL of natural gas is 5% gas to 95% air.

Upper explosive limit (UEL)

The upper explosive limit is the maximum percentage of gas mixed with air which will burn or explode; for example, the UEL of natural gas is 14% gas to 86% air.

Toxicity

Towns gas and processed natural gas have been the only toxic gases marketed for domestic use in Australia as they contain carbon monoxide from the manufacturing process. Although the other gases discussed are not toxic, it must be remembered that with an escape of gas the dangers of fire, explosion and asphyxiation may be present.

Odorants

All gases have odour (usually 2-methyl-2-propanethiol or thiacyclopentane) added before distribution to the customer. Although not all gases use the same odorant, the odour is very similar. The amount of odour added to the gas is sufficient to make a mixture of one-fifth of the LEL detectable by smell; for example, 1% natural gas to 99% air is detectable by smell. The detection of escaped gas by smell indicates only that a mixture of one-fifth of the LEL or more exists; to determine the actual percentage of gas to air a leak detector must be used.

DISTRIBUTION OF NATURAL GAS

The distribution of natural gas commences from the producer's treatment plant where it passes through the transmission pipeline to the city-gate station.

At the city-gate station, the gas changes ownership from the pipeline authority to the gas utility. The volume of gas is measured at this point and then the utility adds an odoriser and controls the pressure to that required in its own trunk mains. The very high pressure of gas in the transmission lines is often sufficient to supply gas for several days to the utility's customers.

Trunk mains from the city gates operate at very high pressures that allow large volumes of gas to be distributed through comparatively small steel pipes. Some customers are supplied directly from the trunk mains, but the majority of customers are supplied from supply mains. The gas passes from the trunk mains into a district regulator that reduces the pressure to that required in the supply main.

The pressure in the supply mains will be in one of the following categories:

Low pressure—up to and including 7 kPa

Medium pressure—over 7 kPa and up to and including 210 kPa

High pressure—over 210 kPa and up to and including 1050 kPa

Low pressure supply mains supply gas direct to the customer. Some utilities use compensating low pressure service regulators, while others supply gas direct to the meter without a service regulator.

Medium pressure supply mains normally supply the customer direct through a medium to low pressure service regulator. However, medium pressure supply mains may also supply gas to low pressure supply mains through a district regulator.

High pressure supply mains either supply the customer direct through a high to low pressure service regulator, or gas is passed through a district regulator and reduced to medium or low pressure.

Figure 1.2 shows a typical natural gas distribution system.

FIG 1.2 Natural gas distribution

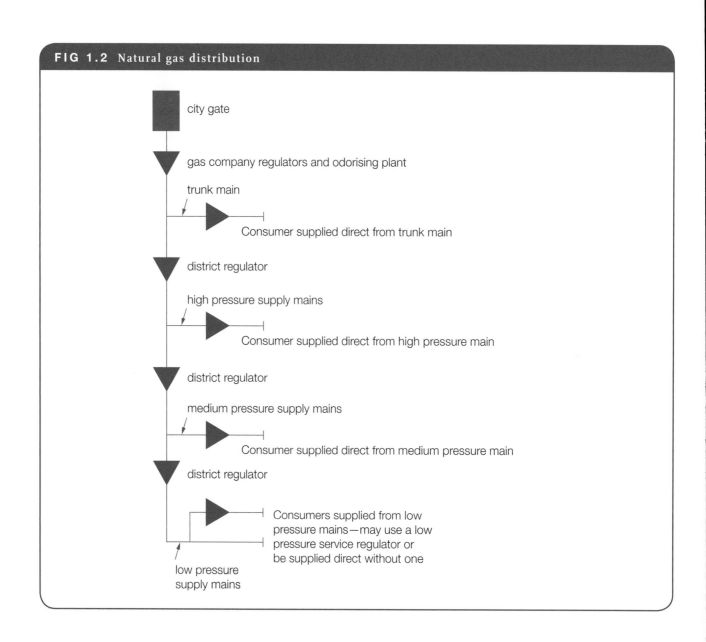

CHAPTER TWO

Pressure and the measurement of pressure

MEASURING PRESSURE AND CONVERSION FACTORS

Pressure

Pressure provides the energy for the distribution of gas and correct operation of burners. The pressures used in the gas industry vary considerably from pressures of up to 3 MPa in transmission lines to very low pressures of less than 1 kPa for the working pressures of some appliances. The base unit of measurement for gas pressure in Australia is kilopascals (kPa), but many appliances, either old or from overseas, have data plates showing the pressure in inches, water gauge (WG), millibars (mb), pounds per square inch (psi), atmospheres (bar) or inches of mercury (inHg). The plumber and gasfitter needs to understand all of these and the gauges that are used to measure them.

Conversion factors

For many years to come there will be a need to convert to the various equivalents in gas pressure depending upon the gauge being used or the method of expressing the pressure recorded.

The following equivalents will be useful when converting pressures:

6.895 kPa	= 1 psi
100 kPa	= 1 bar
100 Pa	= 1 mb
1000 kPa = 1 MPa	= approximately 145 psi
1 atmosphere (sea level)	= 14.73 psi or 101.325 kPa or 1013.25 mb
1 kPa	= 10 mbar

READING AND USING PRESSURE GAUGES

Pressure classification

Low pressure—up to and including 7 kPa
Medium pressure—over 7 kPa and up to and including 210 kPa
High pressure—over 210 kPa and up to and including 1050 kPa

The vast majority of outlet services and appliances are tested and approved to operate within the low pressure classification. To measure these small pressures accurately, a manometer is used.

Reading a manometer

A manometer consists of a U-shaped tube and a measuring scale. The tube is filled with water to the zero level. Both legs of the U-tube are open, allowing atmospheric

pressure to be exerted on both sides; therefore, the water level will be the same in both legs (Fig 2.1). A rubber tube is used to connect one side of the manometer to the gas line. The pressure in the gas line is greater than atmospheric pressure, so the water is forced down on one side and up on the other side. By measuring the height of the column of water being supported by the gas pressure, a very accurate measurement of pressure is recorded (Fig 2.2). The reading is taken from the bottom of the meniscus.

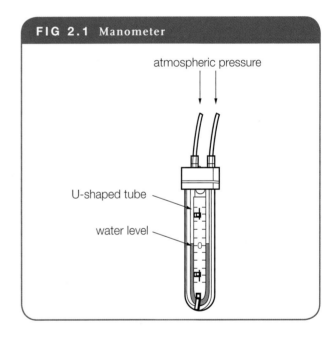

FIG 2.1 Manometer

atmospheric pressure

U-shaped tube

water level

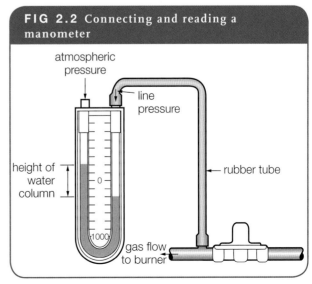

FIG 2.2 Connecting and reading a manometer

atmospheric pressure

line pressure

height of water column

rubber tube

gas flow to burner

The scale used to measure the height of the water column will either be a full-scale indirect reading type or a half-scale direct reading type of manometer.

Indirect reading manometer (full scale)

The method of reading this type of gauge is to read both legs and add them together (Fig 2.3).

Direct reading gauge

The method of reading this type of gauge is to read off one leg only. This is the correct pressure reading providing the water levels were correct at zero before attaching to the gas supply. If water level is not at zero to start with, read both legs, add them together and divide by two.

If difficulty in identifying gauges is experienced, remember you are measuring the height of the column of water being supported by the gas pressure. To identify the type of gauge, measure the height of the column of water with a ruler and use the conversion factors to determine the reading, for example 100 mm equals approximately 1 kPa.

It is most important that the manometer is upright when taking readings because it is the vertical height of the column of water that indicates the pressure.

Mercury-filled manometers

Mercury-filled manometers were used to measure higher pressures accurately. However, they are not used today due to the toxicity of mercury.

Uses of a manometer

A manometer is absolutely essential. Without it appliances cannot be commissioned correctly nor can new installations be tested to the required standards. The maintenance and servicing of appliances also requires the flowing/working pressures to be checked.

Working pressure

Working pressure is the pressure recorded when gas is being used. To set the working pressure of a gas appliance, use the following procedure:

1 Check the data plate to obtain correct working pressure of the appliance.

2 Attach the manometer to a suitable test point on the outlet side of the appliance regulator.

3 Turn the appliance full on; if thermostatically controlled, set to the highest setting.

4 Adjust regulator to correct pressure.

5 Turn on any other appliance in the home that may be used at the same time as the appliance being set. Any drop in pressure indicates an inadequate supply. The cause of the inadequate supply should be identified and corrected.

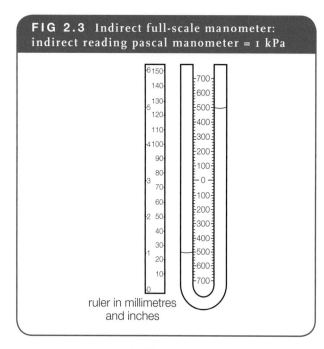

FIG 2.3 Indirect full-scale manometer: indirect reading pascal manometer = 1 kPa

ruler in millimetres and inches

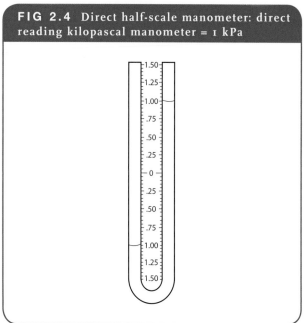

FIG 2.4 Direct half-scale manometer: direct reading kilopascal manometer = 1 kPa

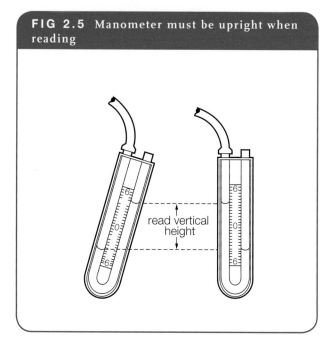

FIG 2.5 Manometer must be upright when reading

read vertical height

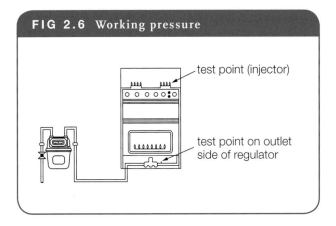

FIG 2.6 Working pressure

test point (injector)

test point on outlet
side of regulator

Static pressure

Static pressure is the pressure recorded when no gas is being used on the outlet side of the meter. Energy is used, moving the gas through the outlet service. If there is no gas being used, there will be no pressure loss. To record static pressure, the manometer can be attached to any pressure point on the outlet side of the meter. To test the static pressure of the system, use the following procedure:

1 Attach the manometer to a suitable point on the outlet service.

2 Turn off all appliances.

3 Take the reading with the main valve on. The reading indicated is the 'lock-up' pressure of the service regulator (i.e. the pressure it shuts off at).

Locating partial blockages

The allowable drop in working pressure from the outlet of the meter to an appliance varies with the metering pressure. Refer to your local authority as to which these are. The allowable drop on propane is 0.25 kPa from the outlet of the regulator to the appliance. By checking the flowing pressure at the meter and at various points on the outlet service, partial blockages can be located. Figure 2.8 shows a typical domestic home installation if the flowing pressure with all appliances on was 1.4 kPa at the meter outlet and tests were taken at each of the three appliances. By analysing the flowing pressure at each appliance, the location of the partial blockage will be clearly indicated.

Example 1
Working pressure at the space heater (F) is 1.35 kPa, a drop of 50 Pa. The working pressure at the cooker (E) is also 1.35 kPa, but the working pressure at the hot water system (D) is 0.8 kPa. Assuming that all appliances are full on and the pressures were taken on the inlet side of the appliance regulator, this would indicate a partial blockage or inadequate sized pipe between C and D.

Bourdon gauge

The Bourdon gauge is used to measure pressures beyond the scope of a manometer filled with water or mercury. The Bourdon gauge indicates the pressure in the line by the pressure exerted in the tube causing a deflection. The amount of deflection in the tube is recorded by the indicator moving around the graduated dial (Fig 2.9).

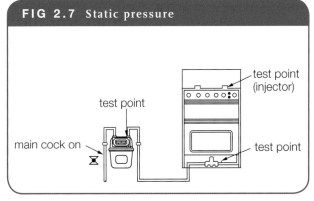

FIG 2.7 Static pressure

test point (injector)

test point

main cock on

test point

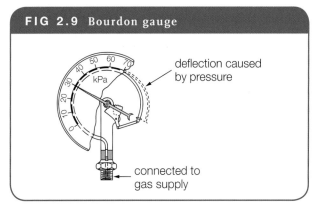

FIG 2.9 Bourdon gauge

deflection caused
by pressure

kPa

connected to
gas supply

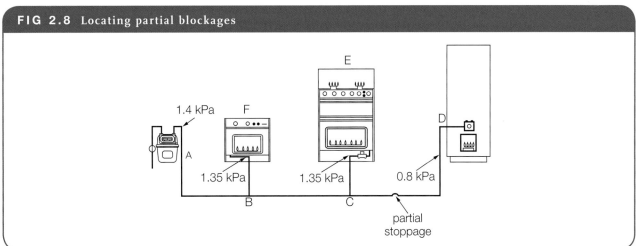

FIG 2.8 Locating partial blockages

E

1.4 kPa   F

D

A

1.35 kPa   1.35 kPa   0.8 kPa

B   C

partial
stoppage

The Bourdon gauge is easily calibrated and should be checked periodically. It is not as accurate as the manometer and should not be used in place of it. If the pressures to be tested are outside the scope of a manometer, check with the local authority before using a Bourdon-type gauge. It is important to choose the correct gauge for the pressure being measured. For example, a suitable gauge for measuring 20 kPa would be one that is calibrated up to 100 kPa rather than 1000 kPa.

PRESSURE RECORDERS

It is sometimes necessary to check the district pressure or the pressure downstream of the meter continuously for 24 hours to ensure that a satisfactory supply is available at all times. A pressure recorder is capable of indicating the pressure at the point to which it is connected for a period of 24 hours or more depending upon the timing mechanism and the charts being used.

The recorder is connected to a suitable point in the pipeline which must be checked. Pressure recorders are normally supplied and fitted by the local authority, where they also ensure that they are necessary. Therefore, the plumber and gasfitter should present all the facts to the local authority if it is felt that a pressure recording is necessary.

TYPES OF PRESSURE

Atmospheric pressure

Atmospheric pressure is exerted by the weight of a column of air approximately 11.25 km high. The pressure in a gas main must be greater than atmospheric pressure, otherwise it would not be able to force gas out of the injector and through the burner because both of them are subject to atmospheric pressure.

Gauge pressure

When a manometer shows a pressure of 1 kPa, the actual pressure in the pipeline is atmospheric pressure plus 1 kPa, but for convenience the pressure is simply recorded as 1 kPa and is known as gauge pressure.

Absolute pressure

Absolute pressure is gauge pressure plus atmospheric pressure and is therefore the total pressure being exerted on a surface.

Effects of altitude on pressure

It is not necessary for the plumber or gasfitter to be able to work out the actual effects of altitude on pressure but he or she should be aware of them. If atmospheric pressure is exerted by the height of the column of air above the surface, then obviously an increase in altitude will lessen the height of the column of air above the surface, thus reducing atmospheric pressure. If the gauge pressure is 1 kPa and the atmospheric pressure at the base of a high-rise building is 101 kPa, then the absolute pressure at this point is 102 kPa. If a reading were taken 60 m higher up the building, the atmospheric pressure would be 101 kPa less the weight of 60 m of air.

The absolute pressure would be 102 kPa less the weight of a 60 m column of gas. The net effect of this is that with gases that are lighter than air, an increase in gauge pressure will take place with an increase in altitude. For example, with natural gas at ground level, the gauge pressure is 1.5 kPa; an increase in altitude of 60 m will cause the gauge pressure reading to increase to 1.8 kPa.

FIG 2.10 Pressure recording chart

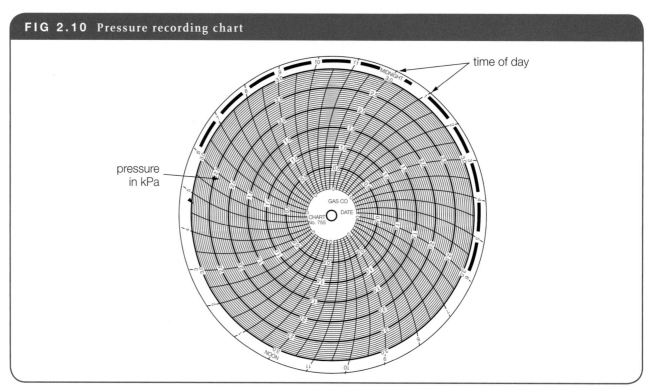

CHAPTER THREE

Principles of combustion and burners

PRINCIPLES AND PRODUCTS OF COMBUSTION

An understanding of combustion is essential whether you are installing or servicing gas appliances. The installer, when commissioning appliances, must be capable of recognising poor combustion and its causes or hazardous conditions may occur. The maintenance or service person must also understand combustion, be capable of identifying causes of incomplete combustion and, when necessary, carry out a suitable combustion test.

Principles of combustion

Combustion is a chemical reaction that takes place at a high temperature and results in the oxidation of the gas. For combustion to take place three ingredients are required:

1 fuel (hydrocarbons)

2 oxygen

3 ignition (heat).

Provided a combustible mixture of gas and air is maintained, the combustion process is self-sustaining once ignition has taken place.

Products of combustion

As we can see from Figure 3.1, three main things come out of the combustion process and are referred to as the products of combustion:

1 heat, which, of course, is all we really wanted

2 hydrogen, which when oxidised becomes water vapour (H_2O)

3 carbon, which when completely oxidised becomes carbon dioxide (CO_2).

FIG 3.1 Combustion process

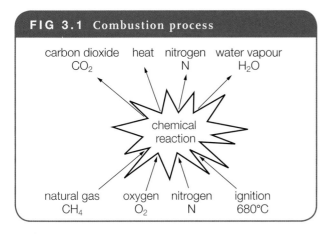

carbon dioxide   heat   nitrogen   water vapour
CO_2 N H_2O

chemical reaction

natural gas   oxygen   nitrogen   ignition
CH_4 O_2 N   680°C

Carbon dioxide is non-toxic and in itself harmless, but if allowed to build up it will exclude oxygen from the combustion process and from the air we breathe.

Water vapour is also harmless, but if it comes into contact with anything below its dew point, it will condensate and this may cause problems.

Nitrogen

The oxygen used in the combustion process is taken out of the air. As air consists of approximately 20% oxygen and 80% nitrogen, what happens to the nitrogen?

Nitrogen is an inert gas and simply passes through the combustion process unchanged except for an increase in temperature. When combustion takes place, a large percentage of the gases that may need to be removed by a flue will consist of nitrogen.

Combustion formulae

As indicated in the characteristics of gases, a certain amount of oxygen is required to oxidise the gas completely, and this varies according to the type of gas being used. For example:

Combustion formula for natural gas.

1 of gas + 2 of oxygen + ignition = heat and products of combustion

or $CH_4 + 2O_2 + 680°C = heat + CO_2 + 2H_2O$

Combustion formula for propane.

1 of gas + 5 of oxygen + ignition = heat + products of combustion

or $C_3H_8 + 5O_2 + 490°C = heat + 3CO_2 + 4H_2O$

The combustion formulae quoted are ideal mixtures with exactly the right amount of oxygen required to burn the gas completely (complete or stoichiometric combustion). To adjust burners to achieve this ideal combustion is not practical, as a burner's performance can deteriorate. So, even if a burner was adjusted exactly to the combustion formula it would in time result in incomplete combustion. To compensate for this, appliances are designed to provide air in excess of that required for complete combustion; this is referred to as excess air. The excess air passes through the combustion process, becomes heated and usually disappears up the flue. For this reason the amount of excess air used must be controlled and kept to a minimum or the efficiency of the appliances will deteriorate. Domestic appliances can be adjusted visually without worrying overmuch about excess air, but large industrial burners may require a flue analysis to determine the volume of excess air being used.

CAUSES OF INCOMPLETE COMBUSTION

1 *Lack of air.* If there is insufficient air and therefore not enough oxygen to oxidise the gas completely, there will be insufficient oxygen to oxidise all the carbon. This deficiency will result in incomplete combustion occurring with the resultant products consisting of carbon monoxide (CO) as well as carbon dioxide (CO_2).

2 *Over-gassing.* Burners are designed to burn a maximum volume of gas and to ensure sufficient air for complete combustion to take place. If a burner is over-gassed there is no guarantee that it will get sufficient air for complete combustion. The net result will be the same as for lack of air.

3 *Blocked or inadequate flue.* Failure to remove the products of combustion may allow a build-up of products that will eventually exclude air from the combustion process starving it of oxygen. Again, the net result will be the same as for lack of air.

4 *Impingement of the inner cone.* Impingement of any solid object into the inner cone will cool some of the fuel below its ignition temperature, thus preventing complete combustion from taking place. This will result in the production of carbon monoxide.

Appliances and burners are designed to ensure that, when fitted and adjusted correctly, complete combustion will take place. By following the manufacturer's instructions, installation codes and the correct commissioning of appliances, complete combustion will take place and this will ensure that the appliance will perform safely and efficiently.

TYPES OF BURNERS

1 *Post-aerated.* (This is sometimes referred to as neat or luminous flames.) Gas is supplied at the burner head with all the air required for combustion coming from around the burner head. This type of burner is no longer used.

2 *Aerated.* Approximately half the air used for combustion is entrained into the burner and mixed with the gas. This air–gas mixture is supplied at the burner head with the remaining air required for combustion coming from around the burner head.

Post-aerated burners

Post-aerated burners are only suitable for fast-burning gases such as towns gas or processed natural gas which are no longer used in Australia. Because the flame was luminous it was originally used for street lighting.

Aerated burners

All types of gases are suitable for aerated burners. Part of the air required for combustion is drawn into the burner and mixed with the gas before combustion takes place. Approximately 50% of the air that is required for combustion is drawn into the burner and is referred to as primary air. The remainder of the air required for combustion is taken from around the burner head and is referred to as secondary air. Combustion takes places considerably more quickly than in a post-aerated burner because it requires less air from around the burner. It gives a smaller flame but at a higher temperature than an equivalent rated post-aerated burner.

Advantages of the aerated burner

1 It can be designed to suit all gases.

2 It requires less room for combustion.

3 Impingement with flame allows efficient transfer of heat without affecting combustion.

Disadvantages of the aerated burner

1 It requires more maintenance than a post-aerated burner.

2 It may light back to the injector.

3 There may be a tendency for the flame to lift off unless some means of flame retention is used.

Figure 3.2 shows the parts of a typical aerated burner.

Operation of an aerated burner

1 The *injector* has two main jobs:

 (a) It is one of the key factors in controlling the volume of gas being delivered at the burner. This is achieved by the correct sizing of the injector orifice.

 (b) The jet of gas issuing from the injector must have sufficient energy to:

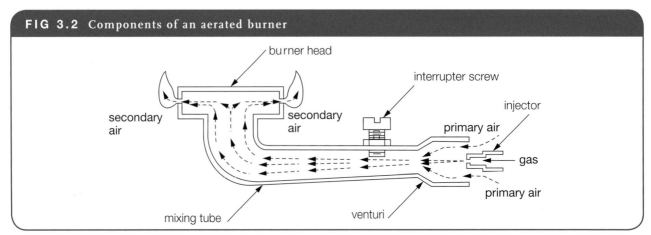

FIG 3.2 Components of an aerated burner

burner head

interrupter screw

injector

secondary air

secondary air

primary air

gas

primary air

mixing tube

venturi

FIG 3.3 Aeration adjustment

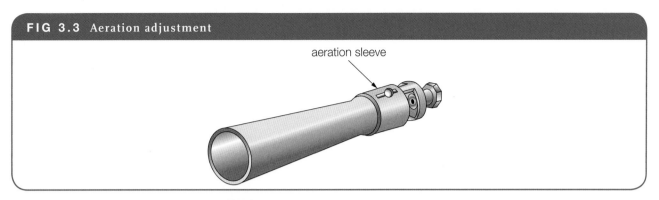

aeration sleeve

FIG 3.4 Aeration adjustment

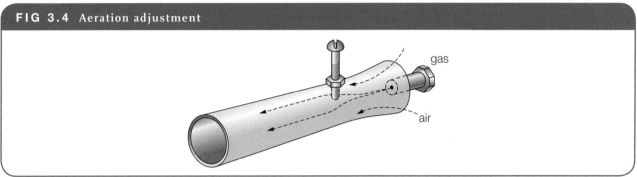

gas

air

(i) provide the energy to draw air into the burner at the primary air port

(ii) provide the energy to push the gas through the burner to the burner port.

2 The *primary airport* is the opening that allows the jet of gas issuing from the injector to draw air into the burner.

3 The *venturi* assists with the drawing in of the primary air and the mixing of the primary air and gas.

4 The *interrupter screw* controls the speed of gas after it leaves the injector by offering a resistance to the gas flow. Gas leaves the injector at speeds of up to 25 km/h. By regulating this speed the amount of primary air drawn in is controlled. An alternative method of controlling the primary aeration is to adjust the size of the primary air opening with a shutter or aeration sleeve.

5 The *mixing tube* allows the air and gas to mix before arriving at the burner port.

6 The *burner port* or *burning area* are the openings in the burner head where combustion takes place.

7 The *burner head* confines combustion to a limited area and determines total flame pattern.

The flame produced by the aerated burner has three distinct zones.

1 The *inner zone,* which is clearly visible, consists of a mixture of unburnt gas and primary air. This inner zone must have clearly defined lines without being jagged or noisy. Correct adjustment is achieved by ensuring the correct alignment and size of the injector, and correct adjustment of the primary air.

2 An *intermediate zone is* where secondary air is mixed with primary air. Combustion begins on the outer edge of the inner zone and is completed at the outer edge of the intermediate zone.

3 The *outer zone,* which is not clearly visible, consists of a mantle of hot gases leaving the flame.

Figure 3.5 illustrates the three zones of an aerated burner.

FIG 3.5 Flame zones of the aerated burner

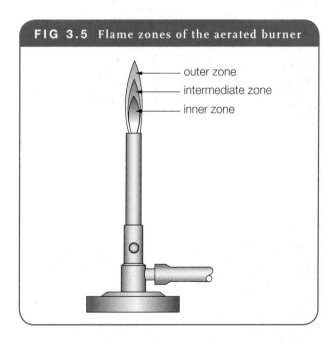

outer zone
intermediate zone
inner zone

Types of aerated burners

There are many different types of aerated burners. Illustrated in Figures 3.6 to 3.9 are some of the more common types. Although the aerated burner comes in a variety of shapes and sizes, its principle of operation does not change.

Burner design

An aerated burner must possess the following qualities:

1 sufficient primary air

2 a steady flame over varying gas inputs

3 ease of control.

The combustion engineer has ensured these qualities by correctly designing the various parts of the burner. It is therefore not necessary for the fitter to know how to design a burner but to be aware of the need to adjust correctly, to understand the principle of operation and to be able to identify the faults that can occur.

FIG 3.6 (Pre-) aerated burner

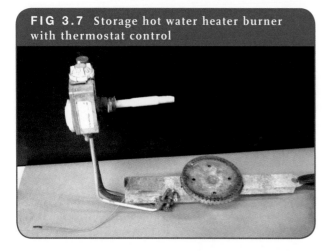

FIG 3.7 Storage hot water heater burner with thermostat control

FIG 3.8 Bunsen burner

FIG 3.9 (a) Cooktop burners

FIG 3.9 (b) Cooktop burner (close-up)

FLAME RETENTION

In order to draw in more primary air, the speed of the gas issuing from the injector had to be increased. This was done by increasing the working pressure of the appliances.

But increasing the working pressure to satisfy the air requirements also increased the problem of the slow-burning properties of natural gas, by delivering the gas at the burner at a faster speed than the flame could consume it. Unless something is done to compensate for this flame, lift-off will occur. Many ways of preventing lift-off are used and are referred to as methods of flame retention. Four common methods are as follows.

1 Increasing the burning area

This would give a larger flame capable of consuming more gas without lifting off. This method, although simple, is very seldom used. The confined space for combustion in most appliances would not allow room for the bigger flame without causing impingement of the inner cone or a lack of secondary air which could cause incomplete combustion (Fig 3.10).

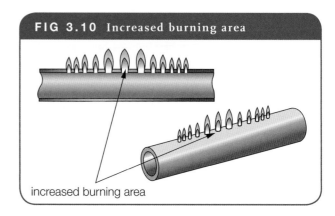

FIG 3.10 Increased burning area

increased burning area

2 Flame retention ports or rings

This method allows the bulk of the air–gas mixture to pass out at the burning area at a high speed, thus giving the flame a tendency to lift-off. Lift-off is prevented by slowing down a small percentage of the gas, which will burn with a stable flame, because the speed at which it is passing through its burning area is slightly slower than the speed at which the flame can consume it. It can be seen in Figures 3.11 and 3.12 that the bulk of the gas passes out of the large burner port or ring, because of its lower resistance to flow, with a smaller amount passing through the smaller

port or ring because of its greater resistance. The small port is referred to as a retention port; it provides a stable flame which prevents the larger flame from lifting off by providing constant re-ignition.

3 Flame retention hardware

This works on a similar principle as the flame retention ports or rings, but instead of using a retention port, contact is made with a small part of the flame. This causes turbulence which slows down a small part of the flame, thus making it stable and preventing the rest of the flame from lifting off (Fig 3.13).

4 Recirculation of the products of combustion

When leaving the flame the products of combustion are above the ignition temperature of the gas. Therefore, if they were to be trapped at the base of the flame, they would have the same effect as a stable flame. Figure 3.14 shows how a burner port can be enlarged in the top section only so that when the burner is lit the flame will impinge on the side of the enlarged section, thus trapping the products and forcing them to pass back through the flame. This recirculation of the products will provide the means of reigniting the flame, thus preventing lift-off. An added bonus for this method is

FIG 3.11 **Flame retention ports**

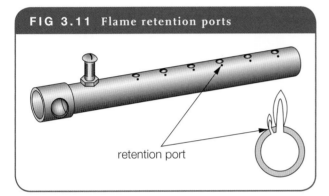

retention port

FIG 3.12 **Retention ring**

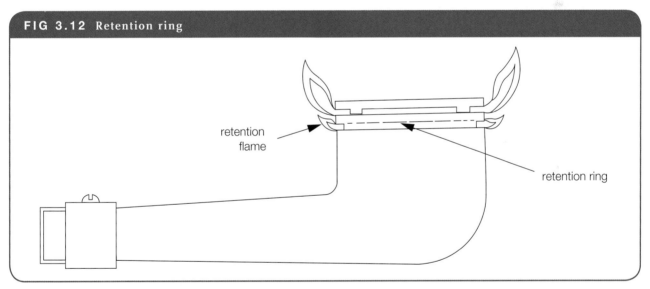

retention
flame

retention ring

FIG 3.13 **Flame retention hardware**

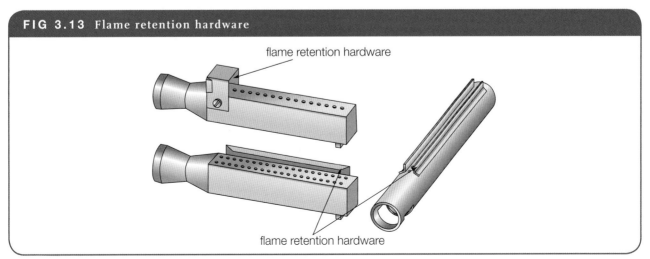

flame retention hardware

flame retention hardware

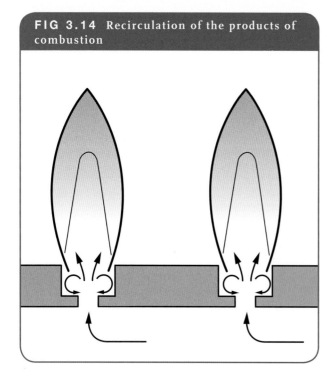

FIG 3.14 Recirculation of the products of combustion

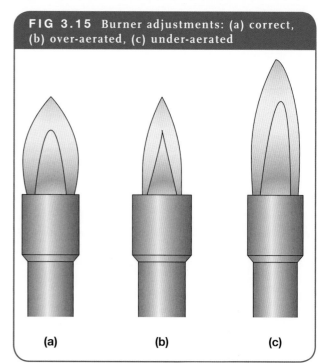

FIG 3.15 Burner adjustments: (a) correct, (b) over-aerated, (c) under-aerated

(a)　　　　(b)　　　　(c)

that the impingement of the flame also stabilises that part of the flame. Although very effective, this method is not often practical for use on domestic burners.

BURNER ADJUSTMENT

An aerated burner should be adjusted when burning the correct maximum volume of gas. Assuming the injector orifice and the type of gas are correct, the plumber and gasfitter must first set the working pressure before adjusting the primary aeration. Aerated burners, unless otherwise stated by the appliance manufacturer, should be adjusted to give the following flame pattern:

1 The inner flame must have straight edges and be dark blue in colour.

2 The tip of the inner flame should be rounded.

3 The intermediate flame should be a purplish haze merging into invisibility (Fig 3.15).

4 The flame should be quiet and steady.

Incorrectly adjusted primary air will result in the following:

1 The inner flame will be very sharply outlined and noisy.

2 The tip of the inner cone is likely to be ragged.

3 The flame may lift-off.

Not enough primary air will result in:

1 an indistinct division between inner and intermediate flame

2 a wall of inner flame likely to bend outwards

3 a yellow tip on the intermediate flame

4 practically no dark blue section.

CHAPTER FOUR

Regulators

The functions of a regulator are:

1 to supply gas at a constant outlet pressure irrespective of varying inlet pressures within design limits

2 where necessary, to provide a means of shutting off the supply when no gas is being used, thus preventing high or medium pressure from affecting a low pressure appliance or meter

3 to provide excess pressure relief where necessary.

All regulators must control the outlet pressure within their pressure range and capacity (function 1 above). Regulators that are required to reduce pressure from the high or medium pressure classification to low pressure must also fulfil functions 2 and 3 listed above.

The movement of the diaphragm and gas valve in most regulators is controlled by variations in downstream pressure. If the downstream pressure increases, it raises the diaphragm and restricts the flow; if the downstream pressure decreases, it lowers the diaphragm and increases the flow. Variations in upstream pressure have a corresponding effect on the downstream pressure, which in turn acts upon the diaphragm

TYPES OF REGULATORS

District regulators are the responsibility of the local gas authority and require highly specialised skills to service and adjust. For this reason, this section will cover only service regulators and the various types of appliance regulator, as follows:

1 service regulators—high to low and medium to low pressure

2 LPG regulator

3 appliance regulator

4 compensating regulator

SERVICE REGULATORS

In districts where the mains are in the high or medium pressure classification, a regulator must be fixed on the inlet side of the meter in order to reduce these pressures to the low pressure classification.

The lower gas valve must be capable of moving up or down against the high or medium inlet pressure. The only available energy to move this valve up and down is the outlet or downstream pressure which is very small in comparison to the inlet pressure. To give this small outlet or downstream pressure the mechanical means of controlling the movement of the valve, a lever is used (Fig 4.1).

To assist the lever with control over the valve, the inlet pressure gas is first passed through a small orifice to reduce the pressure and then allowed to act on a small valve surface. This results in the gas having to pass through a small gap between the valve and the valve seating, which reduces the pressure still further, allowing the low pressure to operate on the bottom side of a large diaphragm. This low pressure spread over a large area is then linked to the end of the lever. The combination of these factors is sufficient to move the small valve up and down against the inlet pressure. In other words, a low pressure exerted over a large surface area coupled with the leverage effect can exert enough force to shut off against the higher incoming pressure exerted over a small surface area.

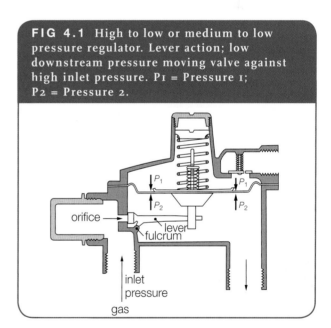

FIG 4.1 High to low or medium to low pressure regulator. Lever action; low downstream pressure moving valve against high inlet pressure. P_1 = Pressure 1; P_2 = Pressure 2.

Operation of a high to low or medium to low pressure service regulator

Refer to Figure 4.2.

1 When an appliance is turned on downstream of the regulator, the pressure under the diaphragm will momentarily drop, lowering the pressure on the lever until it is no longer able to hold the valve against the seating.

2 When the valve moves off the seating, the pressure on the underside of the diaphragm will increase, and it will continue to increase until it is equal to the loading on the top side of the diaphragm. When equilibrium has been attained, no movement of the valve will take place unless

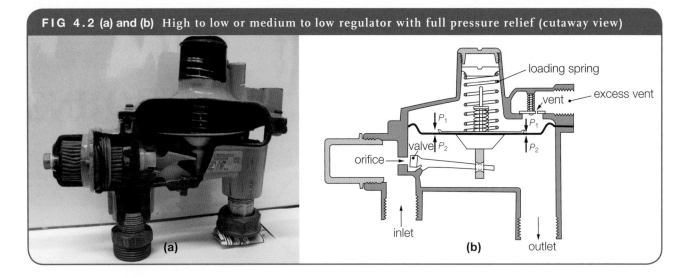

FIG 4.2 (a) and (b) High to low or medium to low regulator with full pressure relief (cutaway view)

(a)

loading spring

vent — excess vent

P_1

valve P_2

P_1

orifice

P_2

inlet

(b)

outlet

there is a change in the inlet pressure or in the demand on the outlet side of the regulator, which will result in either an increase or decrease of downstream pressure.

3 An increase in inlet pressure or a drop in demand downstream will result in a momentary increase in pressure on the bottom side of the diaphragm. This extra pressure, which is linked to the lever, will force the valve to close the gap between the valve and the valve seating. This extra resistance will result in a drop in the downstream pressure until it is equal to the load on the diaphragm (working pressure).

4 A decrease in inlet pressure or an increase in demand downstream would result in the opposite movement to that described in step 3 but with the same end result, equilibrium and resultant working pressure.

5 When all appliances are turned off, the downstream pressure on the underside of the diaphragm will increase. This pressure will continue to increase until the extra pressure which is linked to the lever is sufficient to force the valve hard against the valve seating, shutting

off the supply of gas. This action is described as a 'lock-up' and is necessary to ensure high or medium pressure does not pass through to the meter or appliances when no gas is being used.

The vent

Refer to the operation of the vent on appliance regulators.

Pressure relief section

Refer to Figure 4.3.

When no gas is being used, lock-up cannot be guaranteed due to wear and tear on the service regulator; for example, a worn seating or washer may prevent lock-up taking place. Under these circumstances, the increase in pressure that would pass through to the outlet must be relieved before reaching a dangerous level. The diaphragm, bearing plate and valve are held together in a gas-tight joint by the relief spring. Should the pressure under the diaphragm build up above the tension of the spring, the diaphragm, bearing plate and relief valve would separate allowing the excess pressure to pass out through the excess vent to the atmosphere.

FIG 4.3 (a) Pressure relief section. Pressure relief spring is holding the bearing plate, the diaphragm and the pressure relief valve together in a gas-tight joint. (b) Pressure relief. The excess pressure on the underside of the diaphragm has caused the diaphragm and the bearing plate to separate from the relief valve, allowing the excess pressure to vent to atmosphere. (Detail of Fig 4.2)

bearing plate

pressure relief spring

pressure relief valve

diaphragm

(a)

diaphragm

bearing plate

pressure relief spring

excess pressure

relief valve

(b)

Operation of the excess vent

Refer to Fig 4.4.

For the normal operation of the regulator, the vent must be very small, but should the relief spring operate to relieve excess pressure, the vent needs to be enlarged to allow this excess pressure an uninterrupted path to the atmosphere. The excess vent is sealed by the spring when the regulator is operating normally, but when the relief spring operates, the excess pressure pushes up the excess vent valve, increasing the opening to the atmosphere.

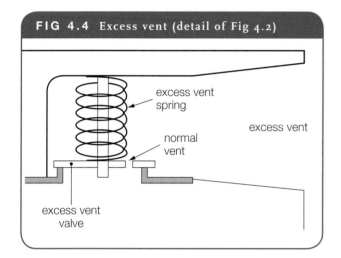

FIG 4.4 Excess vent (detail of Fig 4.2)

LIQUEFIED PETROLEUM GAS REGULATORS

LPG regulators are fitted as close as is practical to the supply source and are often directly coupled to the cylinder. At no time should they be more than one metre away. The pressure in the cylinder can build up to 2,585 kPa and the regulator must be capable of reducing this to the required appliance working pressure of 2.75 kPa. This is achieved in a similar manner to the service regulator by using a fulcrum arm to create enough leverage force for the lower outlet pressure to shut off against a high inlet pressure.

It is also a requirement of the gasfitter to check the lock-up pressure (i.e. the pressure at which the regulator shuts off against the incoming pressure), which should be between 3.0 and 3.5 kPa (usually around 3.25 kPa). As with the service regulator there is an internal pressure relief (keeper plate) to relieve excess pressure in the event of the inlet washer or its seating fouling.

Two stage regulator (LPG)

This type of regulator reduces the cylinder pressure in two stages (Fig 4.5). In effect it is two regulators coupled together with the first regulator reducing the cylinder pressure to 70 kPa and the second reducing the 70 kPa to 2.75 kPa appliance working pressure. This provides a more effective reduction of the cylinder pressure where large volumes are used. To guard against over-pressurisation, a pressure relief valve is incorporated in the second stage regulator diaphragm.

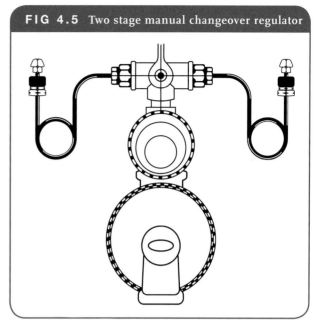

FIG 4.5 Two stage manual changeover regulator

APPLIANCE REGULATORS

If appliances are to burn the correct volume of gas, the pressure must be controlled. It is common in Australia to fit a regulator on the inlet of each appliance for this purpose. This is then referred to as an 'appliance regulator'. Variations in pressure at the appliance are caused by:

1 variations in district pressures caused by the condition of the mains and service and the varying volume of gas required by the customers

2 inadequately sized inlet services, meters and outlet services

3 variation in the number of appliances being used within the household.

On LPG installations there is no appliance regulator and the pressure instead is controlled by the cylinder regulator. LPG appliances either have a test point added to the inlet connection for checking the working pressure or may be tested at an injector nipple and the appliance pressure set to a working pressure of 2.75 kPa where required. Alternatively, the working pressure can be set at the cylinder regulator to 3.0 kPa which allows for 0.25 kPa pressure within the piping system.

Operation of an appliance regulator

The regulator reduces the outlet pressure to the correct level by offering resistance to the flow of gas. For example, if the appliance is designed to operate with a working pressure of 1 kPa and the pressure on the inlet side is 1.25 kPa, the amount of energy used up pushing the gas through the regulator must be equivalent to 0.25 kPa. If the pressure on the inlet side was consistently 1.25 kPa, then all that would be required is a fixed resistance (Fig 4.6).

Figure 4.6 shows that the fixed resistance to gas flow is the gap set between the valve and the valve seating.

Unfortunately, the pressure on the inlet side of the appliance regulator will vary for the reasons previously

seating

outlet

inlet

valve

FIG 4.7 Diaphragm controlled appliance regulator

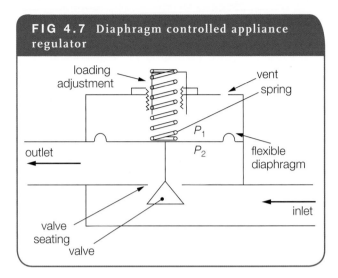

loading adjustment

vent

spring

outlet

P_1

P_2

flexible diaphragm

valve seating

inlet

valve

stated. Therefore, what is required is not a fixed resistance but a variable resistance that can cope with variations in inlet pressure. This variable resistance is achieved by attaching the valve to a flexible diaphragm and asserting a loading on the top side of the diaphragm (Fig 4.7).

All regulators use downstream pressure to control the movement of the diaphragm and valve. An increase in downstream pressure will cause the diaphragm and valve to rise, thereby decreasing the amount gas coming through. A decrease in pressure will allow the diaphragm and valve to move down increasing the amount of gas coming through. This will continue until the downstream pressure is equal to the force exerted by the load placed on the diaphragm. Let us now follow the path of the gas through the appliance regulator assuming the inlet pressure to the appliance regulator is a minimum of 1.13 kPa.

1 Inlet pressure passes through the resistance between the valve and the seating to the underside of the diaphragm and then passes through to the outlet.

2 If at this time the downstream pressure becomes greater than the pressure exerted by the loading on the top side of the diaphragm, the diaphragm and valve will lift.

3 The lifting of the diaphragm and the valve will close the gap between the valve and the seating, thus increasing the resistance by decreasing the space between the valve and its seating, thereby lowering the downstream pressure.

4 When the downstream pressure has been decreased to the point where it is equal to the force exerted on the diaphragm, a state of equilibrium will have been reached and there will be no further movement of the valve.

5 If the pressure on the inlet side of the regulator were to drop, the resistance between the valve and seating would initially be too much, causing a drop in the downstream pressure.

6 A drop in downstream pressure would allow the greater pressure on the diaphragm to push the diaphragm and the valve down, thus lessening the resistance between

the valve and seating as the valve moves away from its seating allowing more gas to pass through.

7 This lesser resistance would allow the downstream pressure to increase until once again an equilibrium is reached. The same would happen if there was a decrease in downstream pressure due to another burner being turned on.

Conversely, an increase in inlet pressure would result in an increase in downstream pressure and the opposite movement of the diaphragm and valve to that described in steps 5, 6 and 7, but with the same end result. Increased downstream pressure can occur due to a burner being turned off.

The vent

The loading above the diaphragm must be constant if the correct outlet pressure is to be maintained. Movement of the diaphragm in the upward direction would compress the air above the diaphragm, thus increasing the loading on it. A dropping of the diaphragm would cause a partial vacuum and have the opposite effect. The vent ensures atmospheric pressure at all times on the top of the diaphragm by allowing the air to move in and out of the upper chamber of the regulator. Rapid changes in inlet pressure may cause the valve to move quickly. If the valve were moved up or down at high speed, the momentum might cause an over-reaction in one direction followed by a reaction in the opposite direction. This would cause the pressure to oscillate. In order to prevent this rapid movement of the valve, the air above the diaphragm is allowed to compress initially when the diaphragm lifts and to bleed out slowly through the vent, thus cushioning the movement of the valve. The size of the vent is therefore very important. An oversized vent would allow the pressure to oscillate; the noise that accompanies this is referred to as 'regulator chatter'.

Loading

The loading of a regulator is achieved by the use of a spring. Originally it was created by the use of a weight placed on top of the diaphragm. However, this had the following disadvantages:

1   The regulator had to be fixed upright in the horizontal plane.

2   It was difficult to adjust.

A spring loading has the following advantages over a weight:

1   The regulator can be fixed in the vertical or horizontal position or, if need be, upside down as long as it is adjusted when in its final position.

2   Spring-loaded regulators are easy to adjust.

3   Spring-loaded regulators are less likely to chatter.

The movement of the diaphragm can change the compression of the spring but this effect is minimised by using a long spring with many leaves. For this reason, the correct spring must be used and never cut under any circumstances.

Accuracy of the appliance regulator

The appliance regulator controls pressure adequately for most appliances, but the main disadvantage of this type of regulator is the effect that varying inlet pressure has on the surface area of the valve (Fig 4.8).

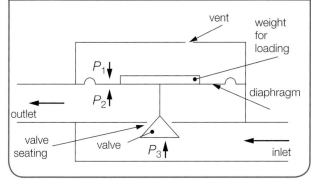

FIG 4.8 Valve effect. As the inlet pressure varies it either increases or decreases outlet pressure, and the regulator has to adjust to compensate for changes in pressure. (P_1 = Pressure 1; P_2 = Pressure 2; P_3 = Pressure 3)

An increase in inlet pressure causes the valve to move up and decrease the downstream pressure. The regulator would then need to correct this by the load on the diaphragm moving the valve back down to increase the downstream pressure to the predetermined setting. A decrease would have the opposite effect. If the inlet pressure varies constantly, the regulator would be constantly adjusting to compensate for the valve effect.

COMPENSATING REGULATORS

Compensating regulators nullify the effect of the variation in pressure on the bottom of the valve.

There are two common types of compensating regulators:

1   Type 1 uses a diaphragm and valve of equal effective area.

2   Type 2 uses a second diaphragm with the same effective area as the valve.

Principle of operation

Both types of compensating regulators work on the same basic principle as the appliance regulator, except for the additional function of compensating for the variation of pressure on the valve, commonly referred to as the valve effect.

The type 1 compensating regulator uses a valve with the same effective area as the diaphragm (Fig 4.9).

FIG 4.9 Simple compensating regulator

loading adjustment
vent
diaphragm
spring
P_1
inlet
P_2
outlet
valve with same effective area as diaphragm

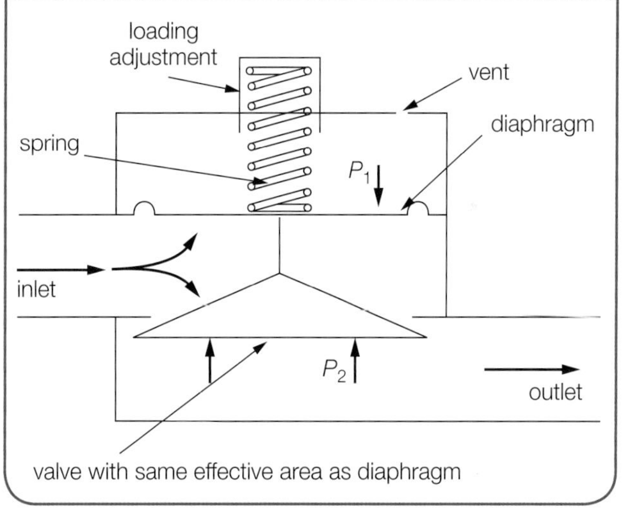

Gas at the inlet pressure pushes up against the diaphragm and down against the valve because they are of the same surface area. The net result is to neutralise one another regardless of inlet pressure. The loading on the diaphragm is therefore only working against the downstream pressure. The equalising of pressure between these two forces ensures a greater degree of accuracy because we no longer have the valve effect regardless of inlet pressure.

The type 2 compensating regulator uses a second (compensating) diaphragm with the same surface area as the valve (Fig 4.10).

Again, the inlet pressure pushes up against the compensating diaphragm and down against the valve, and because they are of the same surface area the net effect is to neutralise one another regardless of inlet pressure. When the gas passes out between the valve and seating, it is then allowed to pass through the channel to the underside of the main diaphragm and to the outlet. The loading on the top of the main diaphragm is only working against the outlet pressure on the bottom side of the diaphragm. The equalising of pressure between these two forces ensures a greater degree of accuracy because we no longer have the valve effect.

FIG 4.10 High to low pressure service regulator

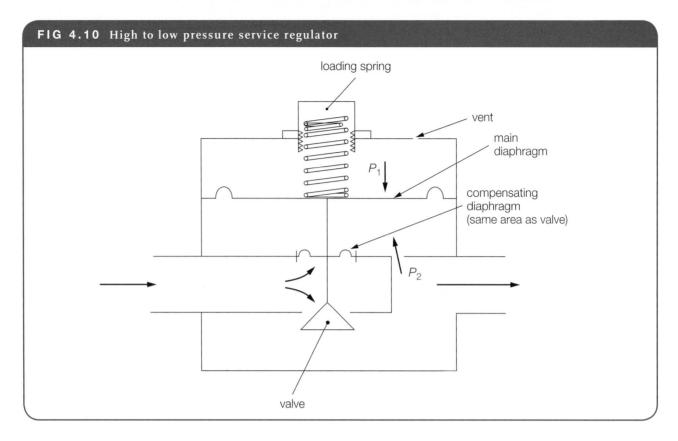

loading spring

vent

main
diaphragm

compensating
diaphragm
(same area as valve)

P_1

P_2

valve

Uses of the compensating regulator

Type 1 is ideal for applying to small appliance regulators. Type 2 is not suitable for small appliances because of the rigidity of the compensating diaphragm, but it is ideal for industrial and commercial appliances and as a service regulator.

Advantages of the compensating regulator

1 It provides more accurate control of pressure.

2 It has a wider operating range within the low pressure classification.

TWO STAGE REGULATION

Two stage regulation uses two separate regulators to take advantage of using higher pressures in the first stage to

decrease pipe sizes in large residential, commercial and/or industrial applications. A typical example of an installation is where a school facility is connected to a LPG bulk storage tank. The first stage regulator reduces the tank pressure to either 70 or 140 kPa, thereby keeping the long runs of first stage piping to a minimal size. The first stage regulator must be installed within one metre of the cylinder (tank) valve and the second stage regulators are usually located at the buildings they serve.

OVER-PRESSURE SHUT-OFF (OPSO)

On larger installations and high metering pressures, over-pressure protection is achieved by using a device which shuts off the gas supply in the event of an over-pressure situation occurring. An over-pressure shut-off (OPSO) is either an integral part of the service regulator or fitted separately. This secondary device senses any over-pressure situation that may cause a problem with the operation of the gas service. When an over-pressure situation occurs, the diaphragm trips a valve, shutting off the gas supply.

The over-pressure device is initially set by pulling back a plunger, allowing the spring setting on top of the diaphragm to engage the interlock mechanism to hold the plunger off its seating, thereby allowing gas to flow through. If the pressure below the diaphragm increases above a predetermined setting, it will raise the diaphragm and release the interlock allowing the plunger spring to return the plunger to its seating, shutting off the gas flow. The plunger needs to be manually reset to restore gas flow.

Over-pressure protection may also be provided by using a monitoring regulator. This is fitted before the service regulator and a pilot line is connected from downstream

FIG 4.11 Typical LPG installation

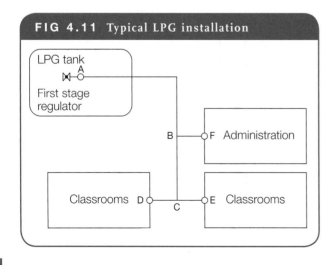

LPG tank

A

First stage
regulator

B — F Administration

Classrooms D — C — E Classrooms

of the service regulator to the underside of the diaphragm on the monitoring regulator. The diaphragm of the monitoring regulator is not in contact with the gas passing through and is therefore not affected by it. The spring setting on the top side of the diaphragm is set higher than the service or controlling regulator and as long as the controlling regulator is operating it is held open. If the controlling regulator malfunctions and over-pressurises, the resultant pressure on the underside of the monitoring regulator diaphragm will cause it to restrict the incoming gas pressure to its predetermined pressure. This situation will be indicated by the pressure gauge located between the two regulators. If the gauge shows the incoming pressure

then the controlling regulator is operating; if it reads the predetermined setting for the monitoring regulator then the controlling regulator has malfunctioned and needs to be repaired or replaced.

UNDER-PRESSURE SHUT-OFF (UPSO)

Under-pressure shut-off (UPSO) is where the gas is shut off when the pressure drops below the minimum pressure required for effective operation of the gas installation. It is used where there is a chance of interruption to the gas supply, such as a coin-operated meter or a pressure-raising device.

FIG 4.12 (a)Over-pressure shut-off (OPSO) with meter and service regulator

FIG 4.12 (c) Over-pressure shut-off (OPSO) detail

FIG 4.12 (b) Over-pressure shut-off (OPSO)

FIG 4.12 (d) Service regulator (detail)

CHAPTER FIVE

Thermostats

FUNCTIONS OF A THERMOSTAT

1 The function of a thermostat is to regulate the temperature of a substance or space by controlling the flow of gas to the burner.

2 A thermostat adds to the convenience, comfort, economy and, in some cases, safety of the appliance.

TYPES OF THERMOSTATS

It is not possible to cover all types of thermostats used in the gas industry, but there are four common types that should be understood by the gasfitter and plumber:

1 rod and tube thermostats

2 liquid expansion thermostats

3 vapour pressure thermostats

4 bimetal thermostats.

ROD AND TUBE THERMOSTATS

Operation of rod and tube thermostats

The basic principle of operation of the rod and tube thermostat is the difference of the coefficient of expansion between metals. The thermostat consists of an outer tube of either brass or copper with an inner rod of invar steel fixed to one end.

One end of the brass tube is anchored, thus ensuring that expansion can only take place in the one direction. Brass expands approximately 18 times faster than invar steel; therefore, when expansion of the brass takes place, the invar steel which is fixed to it will be pulled in the same direction as the expansion of the brass tube (Fig 5.1).

The gas valve follows the movement of the invar steel rod in such a way that when it is cold, the gas valve will be off the seating; but when expansion takes place and the invar steel rod moves, the gas valve will follow this movement. As the gas valve moves closer to the valve seating, it will reach the point where the resistance between the valve and seating is such that it will start reducing the gas flow to the burner. The gas flow gradually reduces to the point where it is only sufficient to maintain the correct temperature. This is then referred to as the holding rate. Thermostats with this type of action are called modulating or graduating thermostats. A minimum gas flow must be maintained to keep the burner alight as there is no means of re-ignition. This minimum rate is known as the bypass rate and it is reached when the valve is hard against the seating and the gas to the burner is passing through the bypass (Fig 5.2).

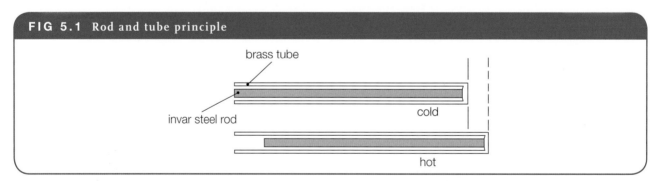

FIG 5.1 Rod and tube principle

brass tube

invar steel rod

cold

hot

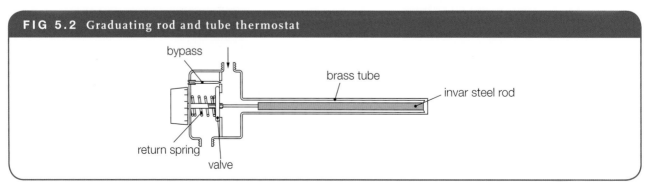

FIG 5.2 Graduating rod and tube thermostat

bypass

brass tube

invar steel rod

return spring

valve

Snap action rod and tube thermostats

Some appliances require the gas to be full on until the set temperature is reached and then shut off. Therefore, cooling must take place before the gas will come back on. This temperature difference between gas off and gas on is referred to as the operating differential of the thermostat. The operating differential can be controlled to suit the type of use, for example, 1°C or less for central heating units or swimming pool heaters and up to 11°C for hot water services (Figures 5.3 and 5.4).

Operation of a rod and tube snap action thermostat

The basic principle of operation is the same as the modulating type, except that the movement of the valve is controlled by a spring steel disc. When the thermostat is cold, the spring steel disc will be in the convex position, holding the gas valve off the seating (Fig 5.3). As the temperature increases, the movement of the invar steel rod will reduce the pressure on the spring steel disc. When the set temperature has been reached, the pressure on the spring steel disc will no longer be sufficient to hold it in the convex position. It will spring to the concave position, allowing the gas valve to shut off (Fig 5.4). As cooling takes place the movement of the invar steel rod will start to apply pressure to the spring steel disc. When sufficient cooling has taken place, the increased pressure on the spring steel disc will cause it to snap into the convex position, pushing the gas valve off the seating to the full on position.

LIQUID EXPANSION THERMOSTATS

Operation of liquid expansion thermostats

Liquids expand when heated and contract on cooling. In a liquid expansion thermostat a bellows is opened and closed by this expansion and contraction. By attaching a gas valve to the bellows, this movement can be used to control gas flow to the burner (Fig 5.5).

The liquid expansion thermostat consists of a phial which is filled with a liquid that has a high expansion rate and is located in the area that is to have its temperature controlled, for example, an oven. Attached to the phial is a capillary tube which relays this expansion to the bellows and on to the gas valve (Fig 5.6).

Liquid expansion thermostats can be either modulating thermostats, as used in an oven, or made into a snap action thermostat by the use of a spring steel disc, such as the ones used on some space heaters. One big advantage of the liquid expansion thermostat is that the sensing phial is remote from the gas valve, allowing for ease of design where the control and sensor cannot be placed together. An example of this is an oven cooker where the sensing phial is placed in the oven and the control is usually on the front of the cooker (stove).

FIG 5.3 Snap action rod and tube thermostat (cold position)

FIG 5.4 Snap action rod and tube thermostat (hot position)

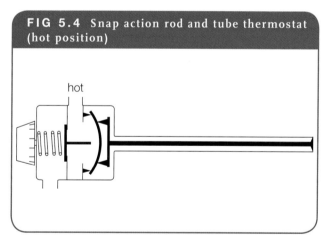

FIG 5.5 Principle of liquid expansion

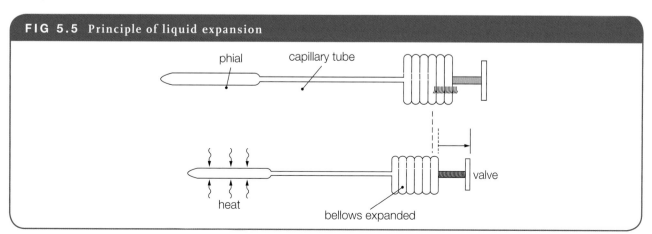

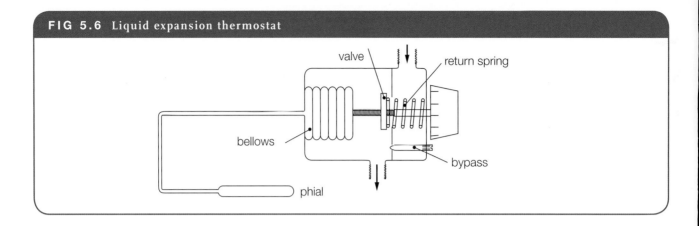

FIG 5.6 Liquid expansion thermostat

VAPOUR PRESSURE THERMOSTATS

The vapour pressure thermostat is used mainly on refrigerators. The sensing phial is fitted to the evaporator and when the cabinet is cooled to the correct temperature the gas rate is reduced to a minimum bypass rate. The burner heats the refrigerant which gives it the pressure for the refrigeration cycle.

Operation of the vapour pressure thermostat

The phial of the thermostat is filled with a gas (e.g. sulphur dioxide) that condenses at approximately 0°C. When the temperature of the cabinet is above the set temperature, the gas pressure in the phial increases. This in turn expands the bellows, thus pushing the gas valve off the seating. As the cabinet cools the gas in the phial condenses, thus reducing the pressure and allowing the bellows to contract and the gas valve to close. A minimum gas rate is maintained through the fixed bypass (Fig 5.7).

Figure 5.7 shows the gas valve in the open position. As the cabinet cools, the valve will move towards the seating. At the set temperature, the gas valve will be in the closed position.

BIMETAL THERMOSTATS

Operation of a bimetal thermostat

There are two types of bimetal thermostats:

1 where two different metals are bonded together, which can be straight, U-shaped or coiled

2 where a single piece of steel is in the form of a spiral and contained in a metal sleeve (Fig 5.8).

In the first type of bimetal thermostat two different metals are bonded together. When an increase in temperature takes place, the resultant expansion will cause the bimetal strip to bend upward or downward depending on the location of the metal with the higher expansion rate. When the two bonded metals form a coil, the coil will wind in an anti-clockwise or clockwise direction, depending on the location of the metal with the higher expansion rate. This movement is used to control the flow of electricity to a solenoid valve.

The second type of bimetal thermostat has a single piece of spring steel in the form of a spiral, contained within a metal sleeve. The spiral is anchored at one end and fixed to a spindle at the other. When expansion takes place, the spiral tightens. This movement rotates the spindle which opens or closes electrical switches controlling the gas supply and the electrical supply to the fan of a central heating unit (Fig 5.8).

Use of bimetal thermostats

Bimetal thermostats are used to control central heating units in the following way:

1 The first type is used to control the fan of a duct air heater and provide overheat protection for the heat exchanger. The function of this thermostat is to operate the two switches contained behind the dial so as to allow the gas to come on first to preheat the heat exchanger, thus preventing the fan from blowing cool air into the rooms. After the burner has been operating for 1–2 minutes, the bimetal thermostat would have rotated sufficiently to have activated the fan switch, allowing the fan to run. When the gas is switched off the fan will continue to run, lowering the temperature of the heat exchanger. This will cause the thermostat to rotate in the

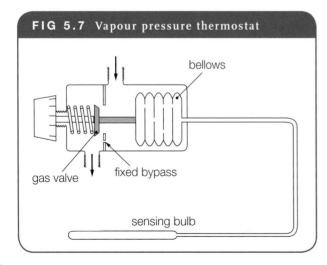

FIG 5.7 Vapour pressure thermostat

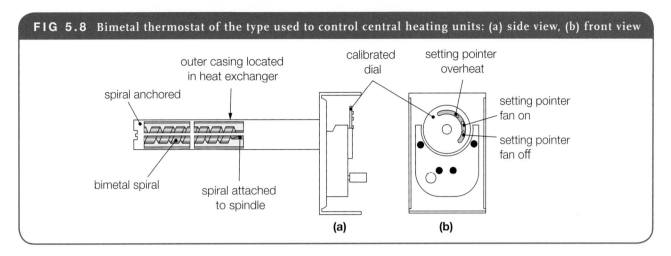

FIG 5.8 Bimetal thermostat of the type used to control central heating units: (a) side view, (b) front view

spiral anchored

outer casing located
in heat exchanger

calibrated
dial

setting pointer
overheat

setting pointer
fan on

setting pointer
fan off

bimetal spiral

spiral attached
to spindle

(a)　　　　**(b)**

opposite direction. At a pre-determined temperature, the fan will be switched off.

If, for any reason, the fan is not passing enough air over the heat exchanger to keep the temperature down to the desired operating temperature, the rise in temperature will cause the thermostat to continue to rotate until it opens the high limit switch, shutting off the gas valve. The fan will continue to run until the heat exchanger cools down again and the burner will come back on.

2　The second types of bimetal thermostats are used in room thermostats and control the gas flow according to the temperature surrounding the thermostat. When the temperature reaches the set point of the thermostat, the gas is shut off. The temperature of the room will now cool until the contacts of the thermostat close, bringing the gas on once again. The room thermostat is therefore not operating at a precise temperature, but remains instead within a certain temperature range called the operating differential. It is most important that in residential premises, this is kept to a minimum, e.g. (1°C—gas off at 22°C, back on at 21°C, operating differential 1°C).

Overshoot

It is very difficult to manufacture a room thermostat that will operate consistently within a small operating differential, because when the room reaches the set point and the thermostat shuts the gas off, the fan will continue to run pushing additional heat into the room until the heat exchanger has cooled to the fan off position. This additional heat into the room will cause the temperature to rise above the set point of the thermostat ('overshoot'), thus increasing the operating differential. The majority of room thermostats compensate for the delivery of this additional heat.

Heat anticipation

The heat anticipator is a small resistor that is used as a heater and is located within the room thermostat, which heats the thermostat slightly above the temperature of the room. This causes the thermostat to shut off the gas

sooner than it would if affected by room temperature only. In other words, the thermostat 'anticipates' the need to shut off the gas early, allowing for the additional heat from the heat exchanger to bring the temperature of the room up to correct setting.

The heat anticipator is wired in series with the gas solenoid valve. Therefore, when the gas is shut off, the heat anticipator is also shut off, allowing the thermostat to cool to actual room temperature. The thermostat will now bring the gas back on immediately the room temperature drops sufficiently to close the contacts. The anticipator is located adjacent to or actually on the bimetal sensor, where the heat produced has a direct effect on the bimetal.

The heat anticipator can be adjusted to control the operating differential of the thermostat. It should be adjusted so that once the room is up to temperature, approximately six gas on–off cycles per hour are achieved (Fig 5.9).

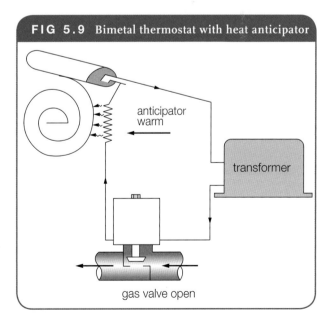

FIG 5.9 Bimetal thermostat with heat anticipator

anticipator
warm

transformer

gas valve open

CHAPTER SIX

Safety devices

FUNCTION AND USES OF SAFETY DEVICES

To prevent unsafe conditions from occurring and adhere to installation codes, safety devices are provided to control the flow of gas to the appliance when the following conditions prevail:

1 Failure of permanent pilots or electronic ignition devices. With many appliances being controlled by thermostats and timers, it is essential that when the supply of gas to the main burner is required, some means of supervising the ignition of that gas must be provided. Should the permanent pilot or electronic ignition fail, the safety device must shut off the supply of gas to prevent a dangerous situation from arising.

2 Excessive high temperatures. Safety devices are used to prevent the overheating of hot water services and heat exchangers.

3 Interrupted supplies or poor pressure. Pressure cut-off devices are used to cut off the supply of gas when the supply has been interrupted or the pressure drops below a predetermined safe or efficient level.

4 Oxygen depletion. The oxygen depletion device is used to cut off the supply of gas to the burner if there is insufficient oxygen to ensure complete combustion.

Thermoelectric flame failure device

Principle of operation

One method of producing electricity is by heating the junction of two dissimilar metals. A gas flame is used to heat the junction, producing approximately 30 millivolts (mV). This heated junction is known as a thermocouple (Fig 6.1). Thermopiles (sometimes called a 'pilot generator') are a manifold of thermocouples in one unit to produce a higher millivolts, generally around 150 to 500 mV. This is done to allow the current generated to operate not only the flame failure and overtemperature cut-out but also to operate thermostat solenoid and other interlocks.

Whenever electricity passes through a conductor, a magnetic field is created around the conductor. By coiling the wire, the strength of this magnetic field can be increased. The strong electromagnet produced has north and south poles and a magnetic field similar to a bar magnet. By coiling the wire of the electromagnet around a soft iron core, the flux density of the magnetic field is increased and the soft iron becomes magnetised, but only while the electrical current is passing through the coil. The electromagnet is used to hold a gas valve off its seating. Should the flame producing the electricity fail, the thermocouple will cool until it is not producing sufficient electricity for the electromagnet to hold the gas valve in the open position. The valve spring will then close the valve (Fig 6.3).

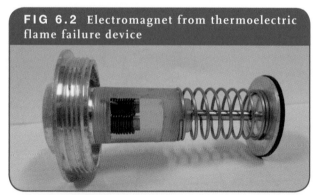

FIG 6.2 Electromagnet from thermoelectric flame failure device

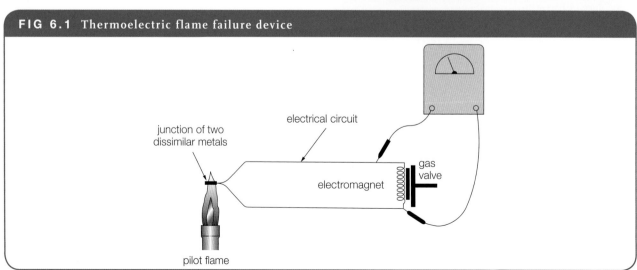

FIG 6.1 Thermoelectric flame failure device

junction of two dissimilar metals

electrical circuit

electromagnet

gas valve

pilot flame

FIG 6.3 Thermoelectric flame failure device—operation: (a) closed position, (b) open position

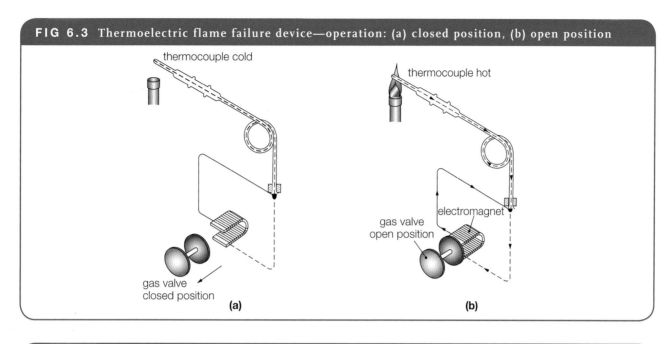

FIG 6.4 Operation of thermoelectric flame failure device: (a) no gas, (b) gas to pilot, (c) gas to pilot and main burner

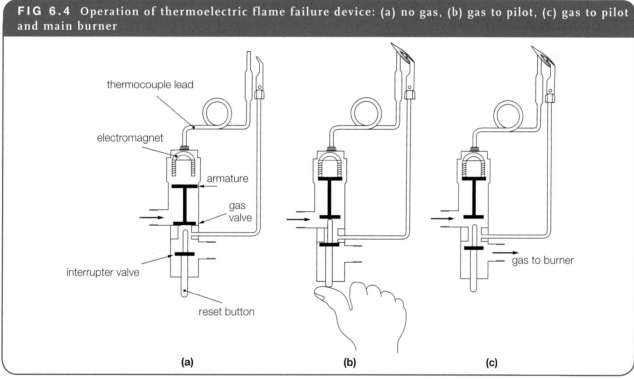

Figure 6.4 shows a typical thermoelectric flame failure device in its three operational positions.

Uses of the thermoelectric flame failure device

The thermoelectric device is used on all types of gas appliances and because of its reliability and cost, it is probably the most common of all flame failure devices.

Servicing of thermoelectric flame failure devices

Check the following points if the thermoelectric device fails:

1 The correct location of the flame which should be on the junction approximately 3 mm from the end of the thermocouple.

2 Excessive bending of the thermocouple can cause a break in the insulation between the inner and outer wire, thus causing a short circuit.

3 Ensure there are no bad contacts between the thermocouple lead and the electromagnet. These contacts must be clean and the joint holding them together should be tightened to hand tight plus one-eighth of a turn.

4 A broken or missing insulating washer can cause a short circuit.

5 Check that the electromagnet coil or contacts are not broken.

Testing of thermoelectric flame failure devices

Select a meter capable of recording accurately a millivolt reading of 1 to 30 mV for thermocouples and up to 500 mV for thermopiles and measure the voltage as shown in Figures 6.5 and 6.6.

FIG 6.5 Testing thermoelectric circuits

Read off 10 scale and multiply by 10 to get correct reading

e.g. needle indicates 3
3 × 10 = 30
= 30mV

Set dial selector to 0.5
(50 μA 0.1 V)

black

over heat

red

electromagnet

A   B

thermocouple

Connect black jack to A or B
Connect red jack to thermocouple outer case

FIG 6.6 Testing suspect thermocouple circuits

1.

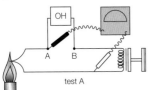

test A

test B

(a) If more than 3 mV difference between test A and B suspect poor contacts to over heat switch

(b) On test B expect at least 12 mV
If more than 12 mV … suspect valve
If less than 12 mV … suspect the thermocouple

2.

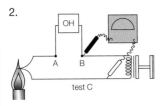
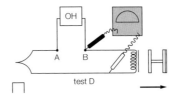

test C

test D

(a) Take reading when thermocouple is hot —C.
(b) Then turn the gas off at gas cock or blow pilot out.
(c) Take reading when you hear the valve drop —D.

The difference should not be less than 5 mV.
e.g. (i) test (a) reading 14 mV test (c) reading 12 mV… suspect valve
e.g. (ii) test (a) reading 10 mV test (c) reading 7 mV… suspect thermocouple
e.g. (iii) test (a) reading 14 mV test (c) reading 8 mV… both OK

Disadvantages of the thermoelectric flame failure device:

1 It requires manual operation.

2 There is a time delay between flame failure and the cooling down of the thermocouple to the point where insufficient electricity is being produced to magnetise the electromagnet to hold the gas valve open. The cooling period can be controlled by varying the mass of the thermocouple. A larger mass takes longer to cool when flame failure occurs and therefore holds the gas valve open longer. For appliances with high rating burners (e.g. an instantaneous hot water service) use a thermocouple with a small mass so it will cool quickly and shut the burner off within 10–15 seconds of flame failure. It is therefore essential when replacing thermocouples that the correct type is used to ensure the burner cuts off within the approved time.

Flame rectification

Principle of operation

Flame ionisation gives a gas flame the capacity to conduct an electrical current. When AC electrical current is passed through the flame from a small electrode to a larger earthed portion of the burner head, considerably more electrical current will flow in one direction than in the other. This partial rectification of the alternating current to direct current can only take place if the current passes through the flame. It cannot be simulated in any other way. The volume of electrical flow through the flame is minute in domestic appliances, usually in the order of five microamps (5 millionths of 1 ampere). The wavelength from this current flow is amplified in the control box and used to operate the solenoid circuit. Any change in the wavelength (e.g. a direct leakage of the alternating current to earth) will be recognised by the control box and the current to the solenoids will be turned off (Fig 6.7).

Uses of flame rectification

Flame rectification is becoming very common in space heating and cooking appliances. The operating sequence on a cooker is as follows.

Electrical current passes through the time switch to the thermostat switch. If both are in the closed position, 240 V AC will pass into the electronic control box. The control box will open the solenoid circuit and, at the same time, provide an electronic spark to ignite the gas at the burner. The control box will allow approximately seven seconds for the gas to ignite; once the gas has ignited, the rectification circuit will be completed through the flame. The control box will recognise this and leave the solenoid circuit open but shut off the ignition circuit. Should the rectification circuit not be proven within the seven second ignition cycle, the control box will go into lock-out, shutting off both the solenoid and ignition circuits (Table 6.1).

Auto cooker troubleshooting

The following checks should be made before trouble-shooting. Checks 3 and 4 should be repeated after all electrical repairs or alterations.

1   Manual shut-off cock in gas line to appliance must be on.

2   Check power to appliance and polarity at power point.

3   Test for short circuit and leakage to earth.

4   Check earth continuity.

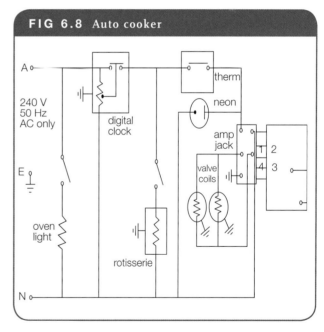

FIG 6.8 Auto cooker

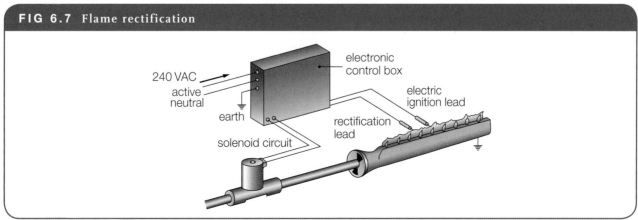

FIG 6.7 Flame rectification

TABLE 6.1 Fault finding chart for flame rectification circuit

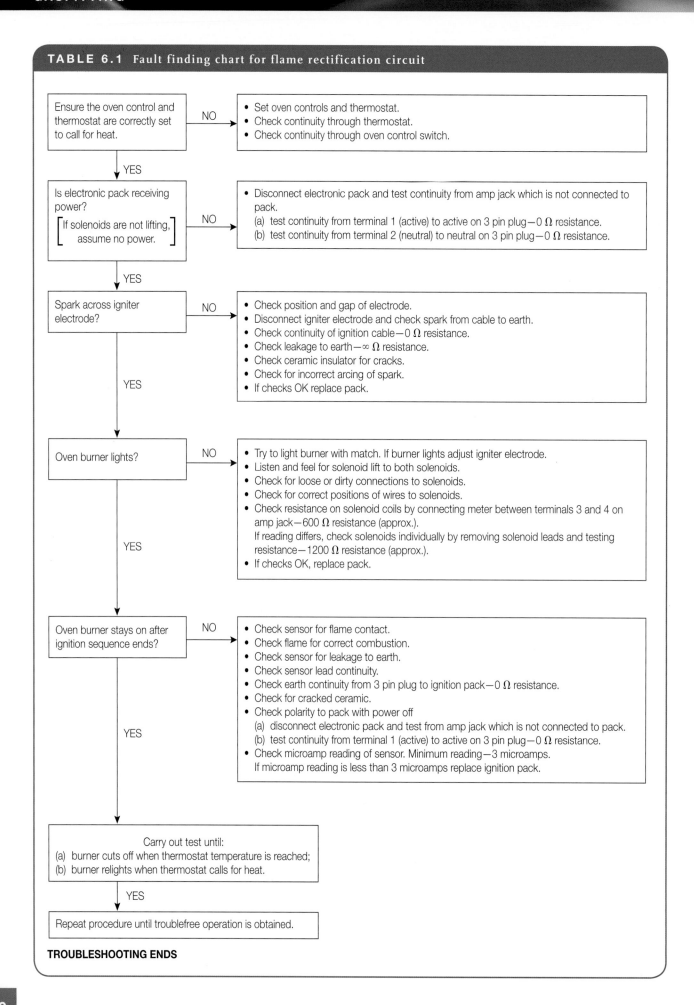

Ensure the oven control and thermostat are correctly set to call for heat.

→ **NO** →
- Set oven controls and thermostat.
- Check continuity through thermostat.
- Check continuity through oven control switch.

↓ **YES**

Is electronic pack receiving power?

[If solenoids are not lifting, assume no power.]

→ **NO** →
- Disconnect electronic pack and test continuity from amp jack which is not connected to pack.
  (a) test continuity from terminal 1 (active) to active on 3 pin plug—0 Ω resistance.
  (b) test continuity from terminal 2 (neutral) to neutral on 3 pin plug—0 Ω resistance.

↓ **YES**

Spark across igniter electrode?

→ **NO** →
- Check position and gap of electrode.
- Disconnect igniter electrode and check spark from cable to earth.
- Check continuity of ignition cable—0 Ω resistance.
- Check leakage to earth—∞ Ω resistance.
- Check ceramic insulator for cracks.
- Check for incorrect arcing of spark.
- If checks OK replace pack.

↓ **YES**

Oven burner lights?

→ **NO** →
- Try to light burner with match. If burner lights adjust igniter electrode.
- Listen and feel for solenoid lift to both solenoids.
- Check for loose or dirty connections to solenoids.
- Check for correct positions of wires to solenoids.
- Check resistance on solenoid coils by connecting meter between terminals 3 and 4 on amp jack—600 Ω resistance (approx.).
  If reading differs, check solenoids individually by removing solenoid leads and testing resistance—1200 Ω resistance (approx.).
- If checks OK, replace pack.

↓ **YES**

Oven burner stays on after ignition sequence ends?

→ **NO** →
- Check sensor for flame contact.
- Check flame for correct combustion.
- Check sensor for leakage to earth.
- Check sensor lead continuity.
- Check earth continuity from 3 pin plug to ignition pack—0 Ω resistance.
- Check for cracked ceramic.
- Check polarity to pack with power off
  (a) disconnect electronic pack and test from amp jack which is not connected to pack.
  (b) test continuity from terminal 1 (active) to active on 3 pin plug—0 Ω resistance.
- Check microamp reading of sensor. Minimum reading—3 microamps.
  If microamp reading is less than 3 microamps replace ignition pack.

↓ **YES**

Carry out test until:
(a) burner cuts off when thermostat temperature is reached;
(b) burner relights when thermostat calls for heat.

↓ **YES**

Repeat procedure until troublefree operation is obtained.

TROUBLESHOOTING ENDS

FIG 6.9 Ultraviolet photoelectric cell

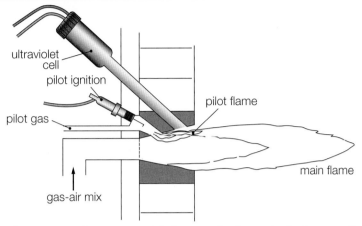

Note: Care must be taken to use the correct sighting method, particularly regarding temperature and in instances where an electronic spark is used, because the sensor is affected by ignition sparks. The sighting method shown in Figure 6.9 is only one method used. Refer to manufacturer's instructions to ensure the use of the correct method.

Ultraviolet photoelectric cell

Principle of operation

Photons are emitted from a gas flame. When they strike the ultraviolet sensing device, they cause a change of wavelength in the current passing through the sensing device. This change in wavelength is amplified and used to control the supply of electricity to the solenoid controlling the gas supply. The ultraviolet sensor is only affected by the small ultraviolet range of the gas flame and not by the infra-red wavelength of either the flame, refractory bricks or process materials. Therefore, it is only in the advent of gas outage that the sensor will shut off the supply of gas (Fig 6.9).

Uses of ultraviolet photoelectric cells

This device is used mainly in industrial applications but may well be used domestically in the future.

Advantages of ultraviolet cell

1 Flame contact is not required.

2 It has a long life.

3 It is suitable for high temperature appliances.

HIGH TEMPERATURE CUT-OUTS

Principle of operation

Sensing devices are used to cut off the supply of gas to prevent overheating of water or heat exchangers. The sensing device is either a fusible link or a heat-activated switch which controls the flow of electricity to the electromagnet of a thermoelectric flame failure device or a solenoid (Figs. 6.10 and 6.11).

FIG 6.10 High temperature cut-outs

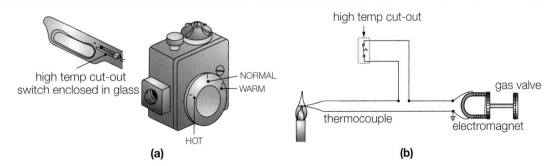

(a)

(b)

Note: The heat-activated switch can be located in the water or heat exchanger that is to be prevented from overheating, and can control the electrical supply to the electromagnet of a solenoid or thermoelectric flame failure device. In some instances, a fusible link is used instead of a heat-activated switch. The disadvantage of a fusible link is it must be replaced. When overheating has taken place, its advantage is that attention is drawn to the overheating problem.

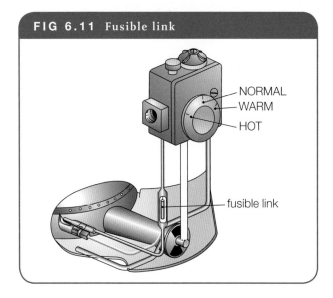

FIG 6.11 Fusible link

NORMAL
WARM
HOT

fusible link

PRESSURE CUT-OFF DEVICES

Principle of operation

If the gas supply is interrupted or falls below a preset level that ensures complete combustion will take place, the pressure cut-off device will shut off the supply of gas to the appliance. The pressure cut-off device operates in the following way:

1 A weighted gas valve is suspended from the bottom side of a flexible diaphragm so that it forms a gas-tight seal over the inlet of the pressure cut-off valve.

2 Inlet pressure is therefore exerted over the small area of the weighted valve. It does not have sufficient energy to lift the valve off its seating against the downward thrust of the weighted valve and the atmospheric pressure that is exerted over the whole area of the diaphragm.

3 When the weighted valve is lifted or pushed off its seating, inlet pressure on the bottom of the diaphragm has sufficient energy to hold up the weighted valve against the downward thrust of its weight.

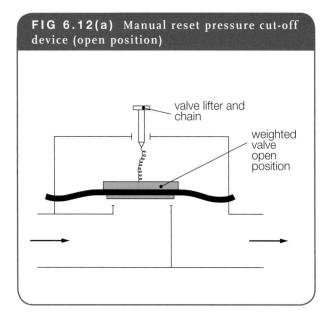

FIG 6.12(a) Manual reset pressure cut-off device (open position)

valve lifter and chain

weighted valve open position

4 Should the inlet pressure drop below the safe set limit, the pressure on the underside of the diaphragm will no longer be able to hold the weighted valve off the seating, and thus the gas will be shut off.

5 The valve must be manually pushed or lifted off its seating when the pressure has been restored, to ensure continuation of the gas supply (Fig 6.12).

Uses of the pressure cut-off device

The pressure cut-off device is used where supplies are likely to be interrupted, for example, prepayment meters which ensure the gas will not be restored until the pressure cut-off device is manually operated.

Pressure cut-off devices are also used on boilers and furnaces to ensure that the minimum pressure requirements for safe and efficient combustion are met.

Pressure switch

Principle of operation

If the gas supply is interrupted or falls below a preset level that ensures complete combustion will take place, the pressure cut-off device will shut off the supply of gas to the appliance. The pressure cut-off device operates in the following way.

A switch is activated by a diaphragm moving in relation to the gas pressure, which in turn cuts the power on the interlock circuit in the event of pressure dropping too low. It can also be used to sense too high a gas pressure and again shut gas off.

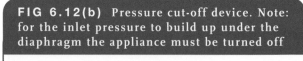

FIG 6.12(b) Pressure cut-off device. Note: for the inlet pressure to build up under the diaphragm the appliance must be turned off

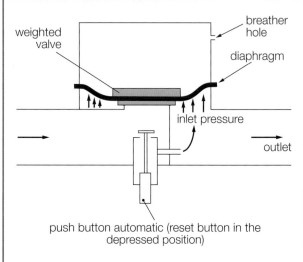

weighted valve

breather hole

diaphragm

inlet pressure

outlet

push button automatic (reset button in the depressed position)

Note: To lift the weighted valve off its seating, the push-button valve is held in the depressed position allowing inlet pressure under the diaphragm. The inlet pressure will lift and hold the valve off the seating.

OXYGEN DEPLETION DEVICES

Principle of operation

To ensure complete combustion, burners are designed to allow for air in excess of that required to oxidise the gas completely. This excess air is necessary so that deterioration in burner performance, which will take place as the burner ages or collects dust and dirt from the atmosphere, will not cause incomplete combustion. The burner will also be affected if the percentage of oxygen in the air drops; a drop in excess of 2% in oxygen content may cause incomplete combustion.

The oxygen depletion device can detect a drop in oxygen content in the room between 1.4 and 2% and will shut off the supply of gas to the appliance within this range of oxygen depletion. The oxygen depletion device heats the thermocouple or thermopile, thus energising the electromagnet and holding the gas valve open. The oxygen depletion device is designed so that its flame speed is equal to the speed of the gas emitting from the burner head, hence it burns with a stable flame. Should the oxygen content of the room drop, the flame speed

will also drop. When the oxygen content of the air is 1.4–2% less than normal, the slower burning speed will cause the flame to lift off and go out. This cuts off the supply of electricity to the electromagnet, which in turn shuts off the gas valve. The oxygen depletion device is, therefore, a very finely tuned burner that remains sensitive to changes in oxygen content. The primary air port is controlled by a bimetal strip that will open if the burner gets hot, so as to maintain the balance between flame speed and the gas emitting from the burner head.

The injector has a ruby orifice that is affected very little by expansion and cannot be drilled out with normal drills, as this would upset the performance of the device. The burner head is made of steatite ceramic that minimises the temperature increase and the improvement in flame retention that an increase in temperature at the burner head would give (Fig 6.13).

Uses of the oxygen depletion device

The oxygen depletion device is used on flueless fires. Consequently, should an over-sized heater or poor ventilation cause a build-up of carbon dioxide and, in turn, a drop in the oxygen content of the room, the drop in oxygen is detected and the heater turned off before it becomes dangerous. Refer to the Australian Standards (AS 5601–2004) for the maximum heater size and other requirements for flueless fires.

Oxygen depletion pilots are commonly referred to as ODS pilots.

Bimetal strips

Bimetal strips are not used in modern appliances. The following information is reference only.

Principle of operation

When two dissimilar metals are bonded together, the difference in expansion rates when they are heated will cause them to bend. The metal with the higher expansion rate will be on the outside, the greater radius resulting from its higher expansion rate (Fig 6.14).

If one end of the bimetal strip is anchored and a valve attached to the other end, when heated by a permanent pilot the movement can be used to control the flow of gas to the main burner. The most common type of bimetal strip is U-shaped (Fig 6.15).

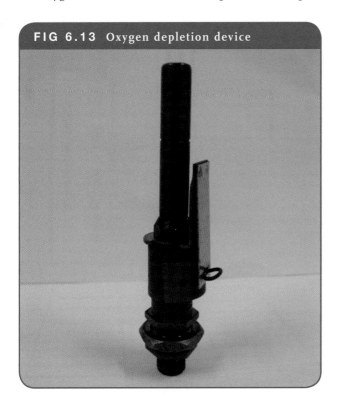

FIG 6.13 Oxygen depletion device

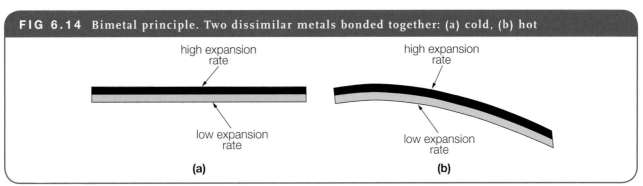

FIG 6.14 Bimetal principle. Two dissimilar metals bonded together: (a) cold, (b) hot

high expansion rate

low expansion rate

(a)

high expansion rate

low expansion rate

(b)

FIG 6.15 Bimetal safety device: (a) cold, (b) hot

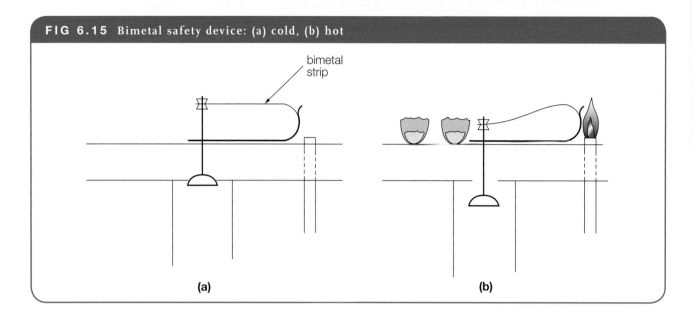

bimetal strip

(a)   (b)

When the permanent pilot heats the bimetal strip, the bending of the bonded metals pushes the gas valve off the seating, allowing gas to the main burner. Should the pilot be extinguished, the bimetal strip will cool and pull the gas valve back onto its seating, thus shutting off the gas supply to the main burner.

Uses of the bimetal strip

The bimetal strip is no longer used as a safety device on modern appliances, but it was quite common on older appliances such as instantaneous and storage hot water services.

Disadvantages of the bimetal safety device

1 Heat from the pilot flame causes metal fatigue. Care must be taken not to allow the main burner flame to impinge on the bimetal strip or it will hasten this fatigue.

2 It does not necessarily fail to the gas off position and therefore is not foolproof.

3 It can be put out of operation by someone leaving the appliance working without a safety device.

CHAPTER SEVEN

Ignition devices

With the introduction of natural gas, there were many changes in ignition devices. Ignition systems, such as battery-operated filaments and flash tubes using either a central pilot, filament or match as the source of ignition, were reliable running on the older and now defunct towns gas, but they proved to be unreliable on natural gas. The main reasons for this were the differing characteristics of these two gases, as follows:

Ignition temperature: natural gas approximately 680° C

Flammability limits: natural gas 5–14% gas to air

As can be seen, natural gas has a high ignition temperature and a relatively narrow flammability range when compared to towns gas. Thus, an ignition device that would allow for these variations was required. Modern ignition devices produce a spark to ignite the burner or pilot light.

PIEZO IGNITION

Principle of operation

When pressure is applied to certain types of crystals, an electrical current is produced. This electrical current is not a continuous flow of electricity but a surge of electricity produced each time the crystals are pressurised. Although the volume (amperage) of electricity is minute, up to 20 000 V can be produced depending on the amount of pressure being applied to the crystals.

The high voltage current is carried by a high tension lead to the electrode. To complete the circuit, it is forced to jump a gap to the burner and then back to its source. When the electrical current jumps the gap the resulting spark ignites the gas.

A piezo ignition device employs a 'hammer' or 'striker' which when activated hammers or strikes the

crystals, thereby compressing them. The electrical current produced travels along the high tension lead to the burner where it jumps the gap and creates a single spark. The spark contains sufficient energy to light the pilot or burner (Fig 7.1).

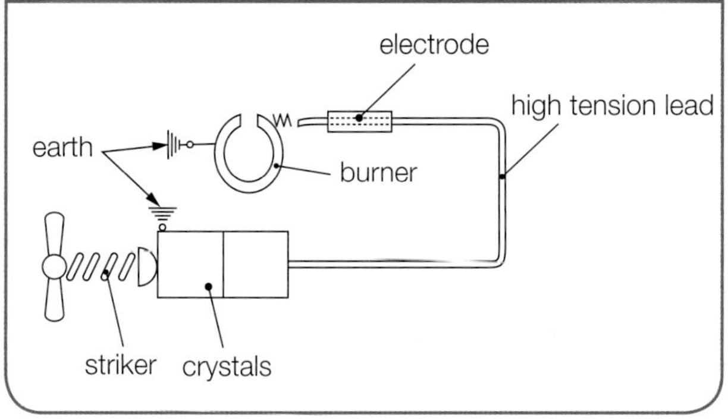

Uses of piezo ignition

A piezo igniter can provide only one spark per compression of the crystals. It is therefore capable of lighting only a single burner and, for this reason, piezo is generally used on single burner appliances, or on cookers for the ignition of the grill and oven burners only.

Fault finding with piezo ignition

The main reasons for the failure to produce a spark is the failure of the mechanical means used to compress the crystals or a break in the electrical circuit.

Check the following points to help identify faults.

1 No spark

(a) Check the crystals are being compressed (mechanical striker).

(b) Check the continuity between the source and the electrode and from the burner or earthing point to the source.

(c) Check there is no leakage to earth. (If checking with a multimeter, the reading should be infinity. *Note:* If leakage is indicated, check the ceramic electrode for cracks.)

(d) Check that the spark gap is not too wide (the gap should be approximately 3 mm).

FIG 7.1 Piezo ignition

electrode

high tension lead

earth

burner

striker   crystals

FIG 7.2 Piezo ignition device

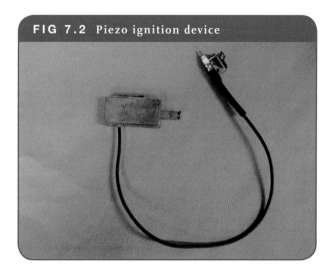

2 Spark but no ignition

    (a) Check the spark gap (set to approximately 3 mm).

    (b) Check that spark is earthing to correct position on burner.

    (c) Check a combustible air–gas mixture exists at the spark location.

PILOT LIGHT IGNITION

Pilot lights originated in the eighteenth century when people kept a small flame continually burning so as to quickly light a fireplace or cooker. Nowadays they are used domestically in natural gas furnaces, gas water heaters and central heating systems.

Safety

A safety cut-off switch is usually included to shut off the supply of gas to the pilot light and heating system if the pilot light goes out. The mechanism uses an electrically operated valve. The two most common ways for the safety switch to detect the pilot lamp are:

1 A photoresistor with electrical circuitry attached detects the light from the flame. If the flame goes out, electric current is generated to close the gas valve.

2 A thermocouple placed in the flame transforms heat to electric current to keep the gas valve open. If the flame goes out, the current is reduced and the valve closes. (See Chapter 6 'Safety Devices' for explanation of thermoelectric flame failure device.)

Energy waste

Concern has been raised over the apparent waste of energy by heating systems using pilot lights, which are estimated to be responsible for up to half of the total energy usage of the system, the average pilot light perhaps using 8–16 Gigajoules/yr. However, since the heat produced by the pilot light is released in the same chamber as the primary burner, the energy goes towards the primary purpose of the system, and so is not in fact wasted.

ELECTRONIC IGNITERS

Electronic ignition devices provide a continuous spark and can provide ignition to a single burner or simultaneously to a number of burners. The electrical source can be either the 240 V AC household supply or a battery, depending on what it was designed to use. The plumber and gasfitter is not expected to become an expert on such electronic devices but when faults do occur he or she should be able to identify whether the fault is internal or external of the electronic box.

If the fault is internal, then the box needs to be replaced. The electronic ignition device produces a pulsating circuit of 10 000 to 15 000 V with a minute amperage and is not therefore capable of being dangerous on the outlet side of the box. However, care must be taken with regard to the input terminals if the source of power is the 240 V AC current, as this can be dangerous.

Principle of operation

When the gas is turned on and the ignition button is pressed, the electronic equipment within the control box will boost the voltage from 240 V or 9 V to 10 000 to 15 000 V. This high voltage current is then transmitted through high tension leads to the electrode, where it is forced to jump a gap to the burner and then back to the ignition box. Providing there is a combustible mixture of air and gas and the spark is of sufficient strength to ignite it, ignition will take place (Fig 7.3).

Fault finding with electronic ignition devices

1 No spark

    (a) Check the source of the electrical supply—battery or mains electricity.

    (b) Check that the spark gap is not too wide (set to 3 mm).

    (c) If supplying ignition to a number of burners and only one or two burners are not getting a spark, swap over the high tension ignition leads at the ignition box. If burners that were igniting before are no longer getting a spark, the fault is in the box. If burners that were igniting before are still igniting after changing the leads, the fault is on the outlet side of the ignition box.

    (d) Check the continuity from the box to the electrode and from the burner or earthing point to the earth connection on the box.

    (e) Check that the high tension leads are not coiled, as this may result in a loss of voltage to the electrode. (Use correct length leads and locate in correct position.)

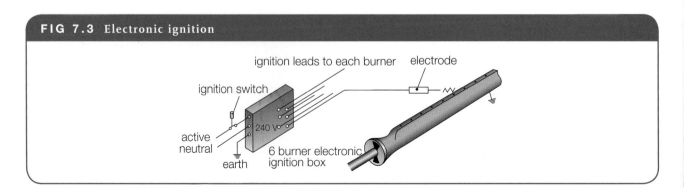

FIG 7.3 Electronic ignition

ignition leads to each burner   electrode

ignition switch

active
neutral

240 V

earth   6 burner electronic
ignition box

2 Spark but no ignition

(a) Check the spark gap (set to approximately 3 mm).

(b) Check that spark is earthing to correct position on burner.

(c) Check that a combustible mixture exists at the spark location.

(d) Check that high tension leads have not been coiled, as this may result in a loss of voltage and a spark of insufficient energy to ignite the gas. (Use the correct length high tension leads and locate them correctly.)

RE-IGNITERS

Apart from providing the means of ignition, re-igniters have the capacity to identify the presence of a flame. The electrode plays a dual role, providing the spark for ignition and then supervising the presence of a flame by acting as a flame rod.

Principle of operation of a re-igniter

When the gas is turned on, a micro-switch is also activated. This allows a spark simultaneously to all burners. When the gas is alight on the burner that was turned on, an electrical current is passed through the flame to the electrode and then back to the re-ignition box. Once the circuit through the flame is recognised by the re-ignition box, the electronic spark is stopped. One advantage of this system is that, should the flame go out for any reason, the re-ignition box will recognise this and start the ignition circuit again until the gas re-ignites.

The electrical current is passed only through the flame. If the gas does not re-ignite on an oven, the electrical circuit will not be complete. This will be recognised by the control box and the gas solenoid controlling the oven will be shut off. This method of flame supervision is known as 'flame conduction' (Fig 7.4).

FLASH TUBE IGNITION

Although very few modern appliances attempt to use this method of ignition, there are still many old appliances and some barbeques with flash tube ignition. So it is necessary to get them to work as effectively as possible when using natural gas appliances.

Flash tube ignition principle of operation

When a burner is turned on, the flash tube charging port injects an air–gas mixture into the flash tube. The air in this mixture has come from the primary aeration intake of the burner, but it is not sufficient to give flash ignition the extra air required for flash ignition is entrained into the flash tube by the velocity of the air–gas mixture emitting from the flash tube charging port.

The mixture of air and gas inside the flash tube has to be very precise. With natural gas, the flash ignition limit is 5.2 to 7% gas to air. Providing the mixture is within this limit, a source of ignition at the ignition end of the flash tube will cause a flashback in the tube, which will light the charging port and the main burner (Fig 7.5).

Fault finding with flash tube ignition

Identifying the reason why a flash tube will not work and then correcting the problem will vary from one flash tube to another and therefore no definite procedure can be given. However, the following points will assist in identifying and correcting problems:

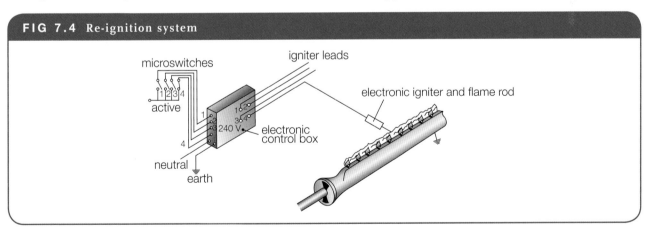

FIG 7.4 Re-ignition system

microswitches
igniter leads
active
1 2 3 4
240 V
electronic control box
4
neutral
earth
electronic igniter and flame rod

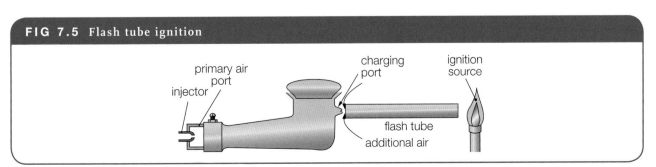

FIG 7.5 Flash tube ignition

primary air port
injector
charging port
ignition source
flash tube
additional air

1 Flash tubes that have worked satisfactorily in the past but have developed problems usually indicates a change in air–gas mixture in the tube. Check the following:

(a) if the flash tube is clear of obstructions

(b) if the flash tube charging port is lighting; if it is lighting, check the reason why it is not tracking to main burner

(c) if too much gas is being injected into the flash tube by over-sized charging port—this problem can usually be identified by the mixture burning at the ignition end of the flash tube

(d) poor alignment of the charging port—gas must be injected straight up the middle of the flash tube

(e) if the flash tube has moved and is either not allowing sufficient air or is allowing too much air to be entrained into flash tube

(f) wrongly adjusted primary aeration of main burner

(g) whether the working pressure of the appliance is correct

(h) if the charging port is partially blocked (undersized).

2 For flash tubes that have never worked check the points described above, but be prepared to make alterations to the flash tube, as follows:

(a) To increase the amount of air being entrained, move or cut flash tube to increase the size of the opening between the charging port and the flash tube.

(b) To decrease the amount of air, move or extend the flash tube.

(c) To decrease the size of the charging port, use permanent materials and re-drill.

(d) To increase the size of the charging port, re-drill.

The mixture in the flash tube **has to be very precise** and therefore any alterations must be done with permanence in mind (e.g. flash tube held rigidly in position). *Note:* Alterations to flash tubes can be time-consuming with no guarantee of success.

FAULT FINDING

Look for faults as for the electronic ignition system, with the addition of the following:

1 When the gas tap activates the micro-switch, the circuit from the micro-switch to the electrode must be to the same burner as the gas supply. Otherwise, the spark will be continuous until the gas tap is turned off (Fig 7.6).

2 There must be flame contact with the electrode.

FIG 7.6 Method of connecting re-ignition system

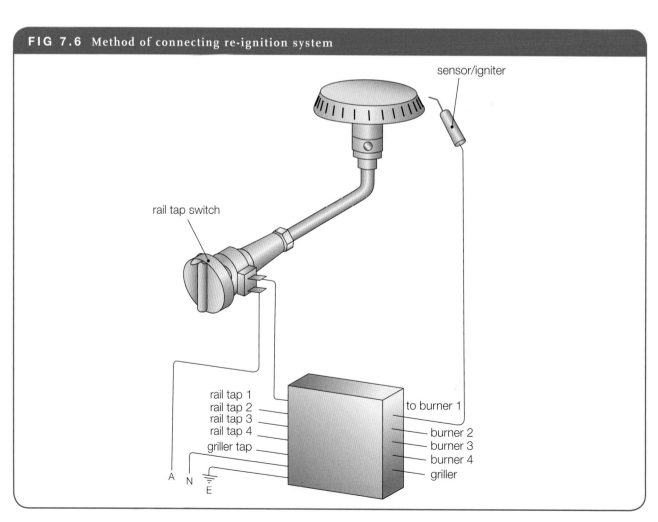

CHAPTER EIGHT

Control devices

THE SOLENOID

The sophisticated electrical and electronic control systems on gas appliances, which increased dramatically in the 1980s, rely almost without exception on the solenoid as the means of controlling the supply of gas to the burner.

Principle of operation of a solenoid

The solenoid consists of an electrical coil and an iron spindle with a gas valve attached to one end. When electricity flows through the coil, it becomes a very strong electromagnet. The electromagnet has sufficient strength to lift the iron spindle and gas valve off its seating. When the electrical supply to the coil is switched off, the electromagnet and the gas valve lose their magnetism and the valve is then returned to the off position by its own weight or with the aid of a spring (Fig 8.1).

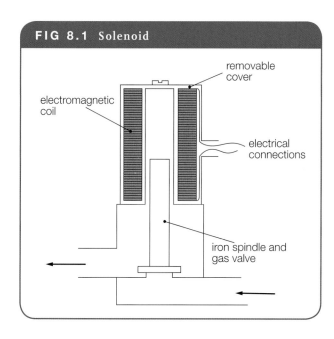

FIG 8.1 Solenoid

removable cover

electromagnetic coil

electrical connections

iron spindle and gas valve

AC solenoids

The alternating current passing through the coil causes the polarity of the electromagnet to change continuously with each change in direction of the current flow. The polarity of the temporarily magnetised iron spindle and valve remains the same. Therefore, because of the laws of magnetic poles—'unlike poles attract' and 'alike poles repel'—the electromagnet continuously attracts and repels the spindle and valve. This continuous attraction and repulsion does not affect the gas flow because the valve is held in the open position. The polarity of the electromagnet causes the spindle and valve to be repelled, but before there is sufficient movement in the spindle to affect the gas flow, the polarity has changed again and attracted the spindle back. This continuous attraction and repulsion causes 'solenoid hum', which can be noisy and difficult to stop.

DC solenoids

When the AC current is rectified to a direct current, the flow of electrical current is in the one direction through the coil and, therefore, the polarity of the electromagnet remains unchanged. With the iron spindle permanently attracted to the electromagnet, there should be no problems with solenoid hum.

Uses of the solenoid

Wherever electrical controls are used to control the supply of gas to a burner, the solenoid is used. In some instances, because of the danger of foreign bodies preventing the valve from shutting off completely, two solenoids are fitted in the gas supply to lessen the chance of gas leaking through.

RELAY VALVES

A relay valve is used in conjunction with safety devices, thermostats, clocks and solenoids to control the flow of gas to the main burner. The principal advantage of a relay valve is that on large diameter pipe supplies, small inexpensive safety devices, thermostats, and so on, can be used to control large burners.

Principle of operation of relay valve

Gas at inlet pressure is allowed to exert a force over the whole underside of a flexible diaphragm. This pressure on the underside of the diaphragm is sufficient to lift the weighted valve, which is attached to the diaphragm, off its seating, thus allowing gas through to the burner.

Gas at inlet pressure also passes through the bypass to the top side of the diaphragm, but this gas passes out through the weep tube without building up pressure on the top side of the diaphragm. The gas that passes through the weep tube is vented near the main burner and is ignited by the main burner.

The control devices are fitted into the weep tube. When the control device closes off the gas flow through the weep tube, the pressure on the top side of the diaphragm builds up. When the pressure on the top side of the diaphragm, plus the weight of the valve, is greater than the pressure on the underside of the diaphragm, the valve will drop onto the seating and close off the supply of gas to the main burner. If a small supply of gas to hold or maintain a temperature is required at the main burner, a valve lifter can be used to hold the valve off its seating; but if an on–off action is required, the valve lifter must be set low so as to allow the valve to seat onto its seating (Fig 8.2).

Common faults with a relay valve

1   Any escape in the weep tube on the inlet side of the control device will prevent or slow down the pressure build-up on the top side of the diaphragm. This will either make the relay valve slow to react or prevent it from shutting off altogether. To test the operation of the relay valve, turn off the weep tube test cock and this should cause the relay valve to close off the supply of gas to the burner quickly.

2   Some relay valves have an adjustable bypass controlling the flow of gas to the top side of the diaphragm. This must be adjusted so that it is not too large an opening or it may cause pressure to build up on the top side of the diaphragm and shut off the gas supply to the burner. Conversely, if the bypass opening is too small when the weep tube control device shuts down, it will take too long for the pressure to build up and therefore be slow in shutting off the gas supply. For this reason, the bypass must be checked and adjusted correctly (Fig 8.3).

COMBINATION CONTROLS

In the past, appliances requiring a control cock, thermostat, regulator and safety device had them fitted separately in the gas line. An alternative to this is to fit a combination control containing all these functions in a single unit. These types of controls have become very common on storage hot water services and space-heating appliances. The thermostat regulator and safety components of a combination control operate separately and independently of each other in the same way as they would if fitted in a line.

One example of a combination control that is used on storage hot water services is Robertshaw Unitrol. Two examples of combination controls used on space-heating appliances are the Minisit and the Honeywell, all of which are discussed below.

Robertshaw Unitrol combination control

This Robertshaw Unitrol combination control consists of a programmed control cock, a rod and tube thermostat, a regulator and a thermoelectric flame failure device with an energy cut-out device incorporated in the circuit. (Refer to Fig 8.4.)

Minisit combination control

The Minisit combination control is used on many types of space heaters and consists of a thermostat to control temperature and act as the appliance control cock, a thermoelectric flame failure device with provision for over-heat protection to the heat exchanger, if required, and a pressure regulator. (Refer to Fig 8.5.)

Honeywell combination control

The Honeywell combination control is used on many types of heating units and consists of a thermoelectric flame failure device, pressure regulator and an operator valve, which is electrically controlled and used in conjunction with a fan limit switch, over-heat switch and a room thermostat.

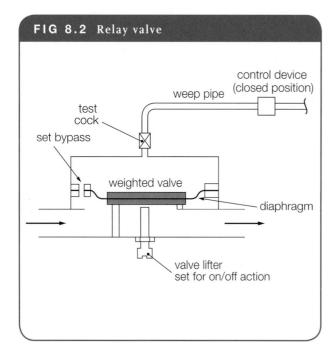

FIG 8.2 Relay valve

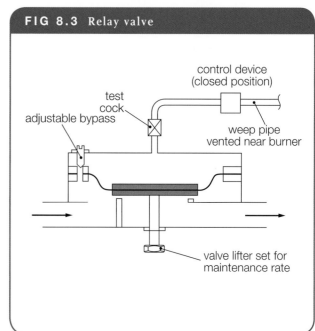

FIG 8.3 Relay valve

FIG 8.4 Exploded view of the Robertshaw Unitrol combination control

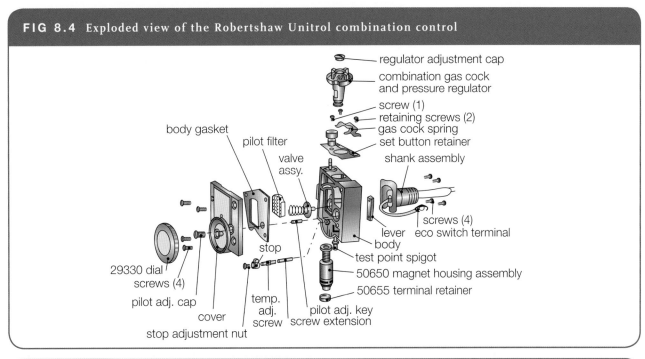

regulator adjustment cap

combination gas cock and pressure regulator

screw (1)
retaining screws (2)
gas cock spring
set button retainer

shank assembly

body gasket

pilot filter

valve assy.

screws (4)
lever   eco switch terminal
body

29330 dial screws (4)

stop

test point spigot
50650 magnet housing assembly
50655 terminal retainer

pilot adj. cap

cover

temp. adj. screw

pilot adj. key
screw extension

stop adjustment nut

FIG 8.5 Minisit control

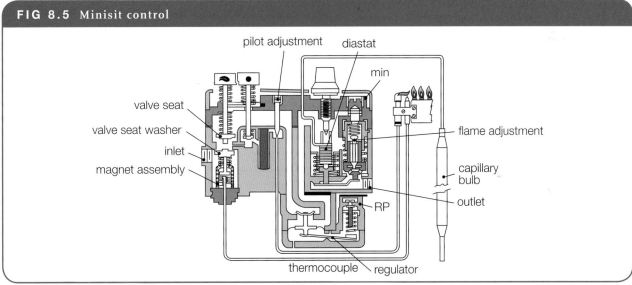

pilot adjustment   diastat

min

valve seat

valve seat washer

inlet

magnet assembly

flame adjustment

capillary bulb

outlet

RP

thermocouple   regulator

FIG 8.6 Honeywell combination control

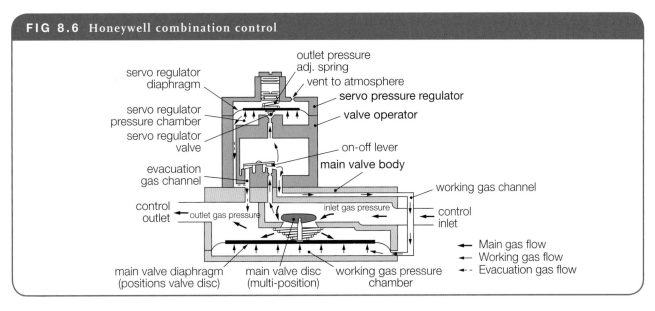

outlet pressure adj. spring

vent to atmosphere

servo regulator diaphragm

servo pressure regulator

valve operator

servo regulator pressure chamber

servo regulator valve

on-off lever

main valve body

evacuation gas channel

working gas channel

control outlet

outlet gas pressure   inlet gas pressure

control inlet

main valve diaphragm (positions valve disc)

main valve disc (multi-position)

working gas pressure chamber

◄━ Main gas flow
◄─ Working gas flow
◄-- Evacuation gas flow

CHAPTER NINE

Fluing and ventilation

NECESSITY FOR A FLUE

The products of complete combustion are heat, water vapour and carbon dioxide. All are considered harmless but if they are allowed to build up the following problems may arise in a room:

1 a hot and damp atmosphere

2 flame vitiation (suffocation), which causes incomplete combustion

3 the release of carbon monoxide from the products of incomplete combustion.

The functions of a flue therefore are to remove the products of combustion in order to prevent the above conditions arising, while assisting with the drawing in of air to the room through the effect of the aeromotive forces created within the flue.

As nitrogen is an inert gas it passes through the combustion process relatively unchanged, apart from being heated.

Factors determining the need for a flue

1 Length of time an appliance is to be used—intermittent or continuous.

2 Megajoule rating of the appliance.

3 Location of the appliance.

Where an appliance is designed to be connected to a flue, it must be connected in accordance with the Australian Gas Association's Gas Installation Code for gas burning appliances and equipment (AS 5601), plus any special local requirements.

OPERATION AND PARTS OF A FLUE

Principle of operation of a natural draught flue

A flue is the means of conveying the products of combustion from the appliance to the flue terminal. The energy for the moving of the products of combustion is generated by the heat contained in these flue gases. The basic principle is that hot gases become less dense and rise (Fig. 9.1). Therefore, providing the products of combustion are hotter than the air surrounding the flue, the products will rise up the flue pipe. This movement, caused by the difference in densities between the hot gases in the flue and the colder air surrounding the flue, is called 'aeromotive force'. The greater the energy difference between the surrounding air

and the flue gases, the greater the aeromotive force.

The aeromotive force conveying the products up the flue is very small and depends on the following factors:

1 the degree of temperature difference between the flue gases and the air surrounding the flue

2 the volume of flue gases contained within the flue. Therefore, if the diameter of two flues were the same, the flue with the greater total height would have more aeromotive force and a greater carrying capacity.

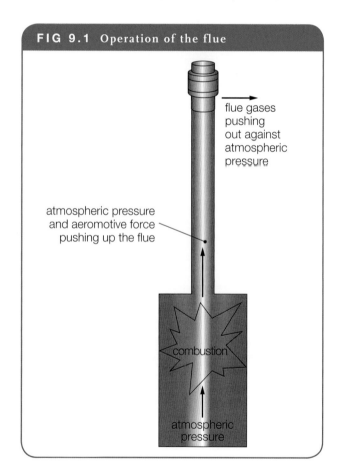

FIG 9.1 Operation of the flue

flue gases pushing out against atmospheric pressure

atmospheric pressure and aeromotive force pushing up the flue

combustion

atmospheric pressure

Parts of a flue and their function

Primary flue

The primary flue is the small section of flue between the heat exchanger and the draught diverter, usually between 50 and 75 mm long. The length of the primary flue must not be increased or it will increase the pull on the flue and lower the efficiency of the appliance.

FIG 9.2 Parts of the flue

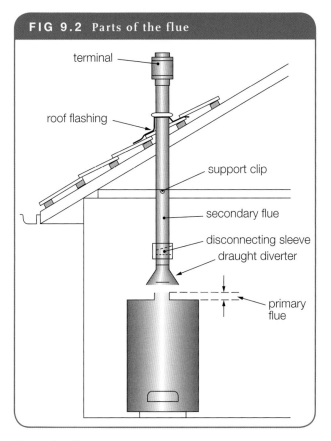

terminal
roof flashing
support clip
secondary flue
disconnecting sleeve
draught diverter
primary flue

Draught diverter

The draught diverter has three important functions:

1 to divert any down draught and the flue gases that would be forced down the flue with it away from the combustion area (Fig 9.3)

2 to dilute the products of combustion by drawing air into the flue—this dilution of the products lowers the dewpoint and lessens the likelihood of condensation forming in the flue (Fig 9.4)

3 to reduce the aeromotive force to the required level by offering resistance to the flow of products.

Secondary flue

The secondary flue is the section of flue from the outlet of the draught diverter to the flue terminal. The installation of this section of the flue is very important. The following points must be observed:

1 Avoid high heat loss conditions such as long lengths of exposed non-insulated flues. Keep the flue inside buildings, or use twin-skin or an insulated flue to avoid high heat losses.

2 Ensure the first 300 mm above the draught diverter are vertical.

3 Where lateral lengths of flue cannot be avoided, ensure a rise of no less than 20 mm/m is observed and lateral lengths do not exceed those indicated as satisfactory in the AS 5601 code.

Terminal

A poorly designed flue terminal can cause down draught or resist the outward flow of the products. When installing a terminal, ensure that only AGA approved terminals are used. An approved terminal will have been tested to ensure the following:

1 It will not divert wind down the flue, thus minimising down draught.

2 It will minimise resistance to the outward flow of products.

3 It will not allow birds to enter the flue.

4 It provides good weathering, thus preventing rain from entering the flue.

5 It is made of non-corrosive materials and is of a reasonable appearance.

FIG 9.3 Draught diverter

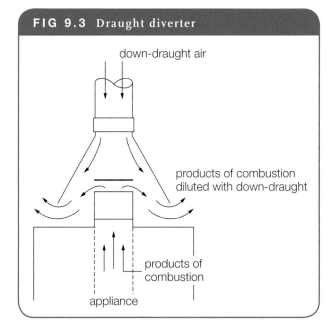

down-draught air
products of combustion diluted with down-draught
products of combustion
appliance

FIG 9.4 Testing operation of a flue

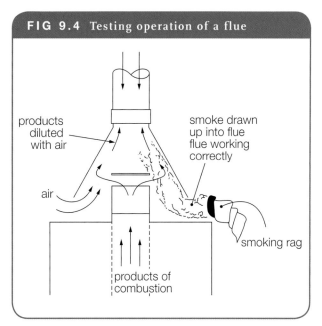

products diluted with air
smoke drawn up into flue flue working correctly
air
smoking rag
products of combustion

Causes of down draught

1   A poorly located flue terminal. Flue terminals must not terminate close to any neighbouring constructions as this may cause wind eddies. Wind may also cause static pressure to build up close to the obstruction, which will cause down draught. Refer to the AS 5601 code for approved flue terminal location.

2   A poorly designed flue terminal.

3   An undersized or poorly designed flue.

4   A negative atmospheric pressure at the point of combustion. This is usually caused by extractor fans extracting air from a room faster than it can be replaced.

Testing the operation of a flue

1   Light up the appliance and set to full gas rate.

2   Leave on for 5 minutes.

3   Use a smoke tester or piece of smoking rag and hold just below the open skirt of the draught diverter.

4   If the smoke is not drawn up the flue, down draught is present or the flue is blocked. (Refer to Fig 9.4.)

FLUE DESIGN

Flue materials

The type of materials to be used on a flue will depend on the following factors:

1   *Temperature of flue gases.* Before installing high temperature industrial flues, consult the local gas authority for approval of the material to be used. Low temperature domestic flues generally use twin-skin metal flue or zinc-coated mild steel flue pipe.

2   *Location of flue.* Flues more than 1 m in length located outside buildings should be of an insulated material.

Materials for the construction of flues should be mechanically robust, resistant to internal and external corrosion, durable and incombustible. For list of approved materials, their application and limitations, refer to the AS 5601 code.

Natural draught single appliance flues

When sizing flues they are generally sized to provide approximately 50% excess air and approximately 100% draught diverter dilution air. Flues required to convey flue products with greater quantities of excess air, dilution air or other products must be designed for the total quantity of flue discharge. Design methods must be based on sound engineering practice.

The tables in the standards show the extent and limitations of natural draught flues relative to the thermal input, height, total length, diameter and other important factors to suit a wide variation in flue configuration.

In determining the correct flue size and configuration for natural draught flue systems, it is essential to consider the heat losses that will occur due to the materials used and the environment in which the flue system will be located.

Since the motive force in a natural draught flue is due to the heat of the flue products, then the ideal conditions are those in which heat losses from the flue system are very low.

Flue materials which are insulated against heat loss (e.g. twin-skin flue) or materials of low thermal conductivity are particularly suitable when the flue is located outdoors or is very long.

Approved non-insulated flue materials when located indoors and not exposed to classified high draught may be classified as 'low heat loss' in applying the flue tables contained in the standards, while the same materials when located outdoors must be classified as 'high heat loss'.

Resistance to flow of the flue products must also be taken into account. The capacities shown in the tables for single flues with laterals make an allowance for two 90° changes of direction in the flue system. If the system requires more than two 90° changes, then a 10% capacity reduction should be made to the table for each additional bend or change (e.g. one additional change, 90% of table capacity; two additional changes, 80% of table capacity).

When flue system dimensions are in between those shown in the tables, flue capacity and configuration may be obtained by calculation.

When determining the flue size for wall furnaces and room heaters, except forced air central heaters, the appliance input should be regarded as being 40% greater than the nominal input on the data plate. For example, a wall furnace having an hourly gas consumption of 40 MJ would need to be sized for 40×1.4 or 56 MJ.

Single appliance flues with laterals and bends that increase the resistance in the flue system in excess of that allowed for in the tables may be designed to minimise flue resistance by increasing the flue diameter from the draught diverter outlet to one size larger.

This will increase the flue system capacity by approximately 60% of the difference between the capacity of a flue system, designed on the basis of the actual appliance or draught diverter flue outlet size, and the capacity of a similar flue system one size larger.

Any further increase in flue size is not recommended as it will not have a similar corresponding effect.

Procedures for using flue tables to size single natural draught flues

Having determined the type of flue material and its location in regard to heat loss, the appropriate flue table may then be used.

The procedures for use of flue tables for single natural draught flues, whether for low heat loss or high heat loss, are identical.

Step 1   Determine the total flue height of the system and length of lateral (Fig 9.5).

Step 2   Read down the total flue height column at the left of the appropriate table for low or high heat loss situations until a height equal to the height of the flue system or nearest below is listed.

Step 3   Select the horizontal row for the appropriate length of lateral (zero for straight vertical flue systems). Select the actual length of lateral or nearest above.

FIG 9.5 Natural draught flue

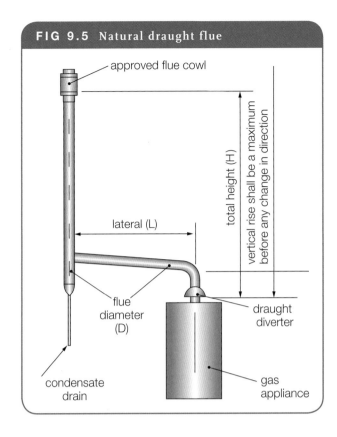

approved flue cowl

total height (H)

vertical rise shall be a maximum before any change in direction

lateral (L)

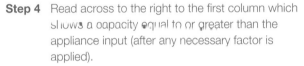

flue diameter (D)

draught diverter

condensate drain

gas appliance

FIG 9.6 Flue sizing

0.5 m

3.5 m

galvanised metal flue

located inside building

125 mm outlet

130 MJ

appliance

Step 4 Read across to the right to the first column which shows a capacity equal to or greater than the appliance input (after any necessary factor is applied).

Step 5 If the flue diameter shown at the top of the column containing the appliance input (or corrected input) is equal to or larger than the appliance flue outlet, use the flue diameter in the table.

Step 6 Where the flue diameter in the table for individual flues indicates a flue diameter less than the appliance or draught diverter flue outlet size, the smaller diameter may be used only when:

(a) the flue height is 3 m or greater;

(b) flues 300 mm in diameter or less are not reduced by more than one size (200 to 175 mm is a one-size reduction);

(c) flues exceeding 300 mm in diameter are not reduced more than two sizes (600 to 560 mm is a two-size reduction).

Under no circumstances shall a 75 mm flue be connected to an appliance having a 100 mm flue outlet.
Note: The maximum allowable heat input to flues may be varied with the approval of the authority.

Example 1
A storage hot water service is to be installed with a flue configuration as shown in Figure 9.6.

Location of flue: inside building except the 500 mm through the roof. Flue material: galvanised metal flue. Draught diverter flue outlet size: 125 mm. Appliance input: 130 MJ/ h.

Step 1 *Use the table for 'Low Heat Loss Situation' from AS 5601.*

Step 2 Under the left-hand column headed 'Total height of flue', select a height equal to 3.5 m or the nearest which is less than that height.

Step 3 Under the column headed 'Length of lateral', select a lateral length within the 3 m total height column equal to or greater than 0.5 m.

Step 4 Read across to the right of the 3 m height and 0.6 m lateral column until an MJ rating equal to or greater than 130 MJ is listed.

Step 5 The third column indicates a flue with a capacity of 136 MJ. The top column vertically above this figure indicates a flue diameter of 125 mm is required.

Example 2
A storage hot water service is to be installed with a flue configuration as shown in Figure 9.7.

Location of flue: outside building. Flue material: metal twin skin. Draught diverter flue outlet size: 150 mm. Appliance input: 140 MJ/h.

Step 1 *Use the individual appliance flue table 'Low Heat Loss Situation' from the Standards.*

Step 2 Under the column headed 'Total height of flue' select a height equal to 4.5 m or the nearest which is less than that height.

Step 3 Add the two lateral lengths together. Under the column headed 'Lateral length', within the 4.5 m total height column, select a lateral length equal to or greater than 1.3 m.

Step 4 Read across to the right of the 4.5 m height and 1.5 lateral column until an MJ rating equal to or greater than 140 MJ is listed.

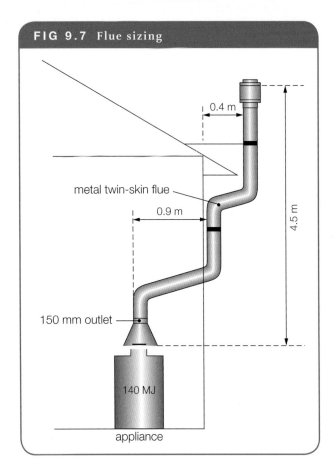

FIG 9.7 Flue sizing

0.4 m

metal twin-skin flue

0.9 m

4.5 m

150 mm outlet

140 MJ

appliance

Step 1 *Use the individual appliance flue table 'Low Heat Loss Situation'.*

Step 2 In the 'total height' column select the height equal to or just less than the total height of the flue.

Step 3 As the flue indicated in Figure 9.8 is vertical, use the 0.0 column under column headed 'Length of lateral'.

Step 4 Read across to the right of the 4.5 m total height column and 0.0 length of lateral column until an MJ rating equal to or greater than the appliance input is listed.

Step 5 The second column indicates 118 MJ; the top figure vertically above this figure indicates a flue diameter of 100 mm is required.

Step 6 When the flue diameter tables indicate a flue diameter less than the draught diverter flue outlet size, the smaller diameter may be used only when:

(a) the flue height is 3 m or greater;

(b) flues 300 mm in diameter are not reduced by more than one size;

(c) flues exceeding 300 mm in diameter are not reduced by more than two sizes.

Note: No reduction is allowed if draught diverter outlet size is 100 mm or less.

Applying the four rules listed above, the appliance indicated in Figure 9.8 may use a 100 mm flue as indicated in the individual appliance flue table.

Step 5 The third column indicates a flue with a capacity of 154 MJ. The top column vertically above this figure indicates a flue diameter of 125 mm is required.

Note: A 125 mm flue with a total height of 4.5 m with a single lateral of 1.3 m would have a capacity of 154 MJ, but the flue configuration shown in Figure 9.7 has two extra bends. The extra resistance of each extra bend reduces the capacity of the flue by 10%.

Step 6 Reduce the 154 MJ capacity by 10% for each bend.

154 less (15.4 × 2) = 123.2MJ

Therefore, the capacity of a 125 mm flue with the two extra bends is 123.2 MJ.

Step 7 Select the next MJ rating to the right of the 154 which indicates 231 MJ with a reduced capacity of 231 less (23.1 × 2) = 184.4 MJ. The top column vertically above this figure indicates a flue diameter of 150 mm is required.

Example 3

An instantaneous hot water service is to be installed with a flue configuration as shown in Figure 9.8.

Location of flue: inside building except the 600 mm through the roof. Flue material: metal twin skin. Draught diverter flue outlet size: 125 mm. Appliance input: 110 MJ/h.

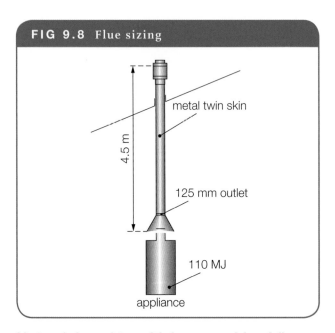

FIG 9.8 Flue sizing

4.5 m

metal twin skin

125 mm outlet

110 MJ

appliance

Natural draught multiple or combined flues

Multiple or combined flues in which two or more appliances are connected to a common vertical flue stack should be designed on the same basic principles for heat loss and flow as with single appliance flue systems. There are, however, a number of important additional matters which must be applied when designing common flues.

The most critical operating condition is when only one appliance is in use, particularly if that appliance input is low compared with other appliances connected to the same common flue.

When a common flue has a manifold or lateral at the base, the flue design must allow for additional resistance to flow due to the change of direction. The (L) lines in common flue table include an allowance for this increased resistance.

Manifolds shall not exceed 50% of the total flue height or 3 m, whichever is the lesser. When a number of appliances are installed in parallel and are designed to operate only simultaneously, and not independently of each other, the manifold and vertical flue may be designed as an individual or single appliance flue, using the individual appliance flue table. The common manifold should then be designed as a lateral length.

Satisfactory performance of common flue systems depends also on careful design of the flue connector, that is, the part of the flue system connecting the individual appliances from the draught diverter outlet to the common flue (Figs 9.9 and 9.10).

The flue connector configuration in diameter, lateral, rise and total length is of major importance, not only to prevent spillage from the appliance draught diverter but to contribute to the correct performance of the common flue.

The flue from the first or lowest appliance connector to the common flue may be designed as a single flue terminating at the first interconnection or tee.

The other appliances joining the common flue at upper floor levels are designed using the common flue tables.

In applying the tables, the total flue height is the rise in the flue connector plus the vertical height between the connection to the common flue and the next connection above (Fig 9.9). The top floor appliance has a total flue height, which is the rise in the flue connector plus the vertical height from the connection with the common flue

to the terminal of the flue. Consideration should be given to providing a separate flue for the top appliance if its total height is inadequate. Where the common flue is more than seven times the area of the flue connector, the rise of the flue connector must be increased by 300 mm more than that shown in the tables.

Example

Water heaters are to be installed on each of four levels in a building. The height between floors is 3 m and each appliance has a 100 mm flue outlet and an hourly gas consumption of 50 MJ.

Length of the lateral in the breeching is 600 mm.

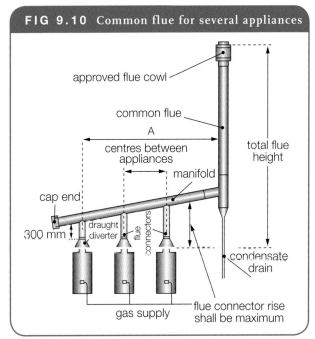

FIG 9.10 Common flue for several appliances

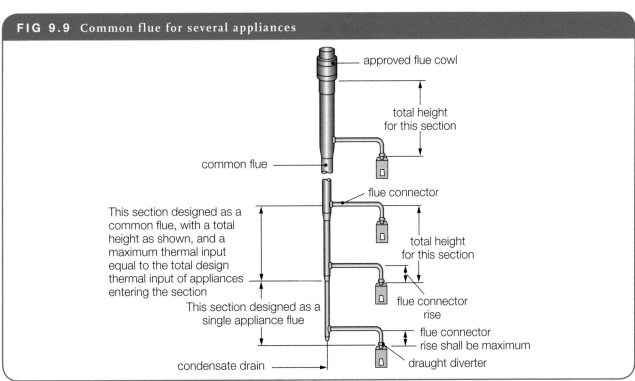

FIG 9.9 Common flue for several appliances

Step 1 The lowest appliance flue may be designed using the single flue for a flue with 3 m total height and 0.6 m lateral; a 100 mm flue diameter has a capacity for hourly gas consumptions up to 85 MJ, which is above that required (i.e. 50 MJ).

Step 2 The tee connection to receive the second appliance and the next section of common flue must have capacity to serve the two appliances (i.e. 100 MJ), but first the flue connector size should be determined. From the flue connector pipe table under least total height, locate 3m. Now reading across to the right, we find with 0.3 m connector rise, a 100 mm flue has capacity for 53 MJ which is adequate.

Step 3 The common flue size to carry 100 MJ is next determined. From the common flue under least total height, locate 3 m. Because the common flue is without change of direction and the appliances are individually attached, Type V applies. Now reading across to the right we find a 125 mm common flue is satisfactory up to 131 MJ.

Step 4 The third appliance is now added to the common flue which then requires capacity for 150 MJ. Two alternatives may now be adopted:

    (a) to design the third section of common flue using the same total height between connections as previously (i.e. 3 m) on the assumption that the top floor appliance will be connected to the common flue; or

    (b) to design on the basis that the top floor appliance will not be joined to the common flue but flued separately. This provides an increase in the total flue height above the third appliance. Assume that this is now 6 m.

Reading the common flue table:

    (a) Under total flue height 3 m type V, reading across the table a 150 mm common flue would be required having capacity up to 188 MJ.

    (b) Under total flue height 6 m, reading across the table, 125 mm common flue would be suitable having a capacity up to 169 MJ.

This illustrates the increase in flue capacity through additional total height which increases the draw of the flue. The choice may be made on grounds of economy and availability of space.

Another method of designing flues for multistorey buildings is to provide an oversize common flue or duct of constant diameter for its total length, and design the breeching flue as individual flues.

They are then classified as self-venting. The common flue acts as a duct for the conveyance of flue products but does not necessarily contribute to satisfactory draught in the flue connectors.

The common flue tables apply when the individual draught diverter outlets from appliances connected to a common flue are within range of the common flue—maximum draught diverter table size. Note that the size of appliance is determined by the draught diverter diameter.

If the largest draught diverter exceeds the range in the common flue—maximum draught diverter table size, then increase the flue connector rise by 300 mm in excess of that shown in the flue connector pipe table.

The flue connector pipe table allows for two 90° changes of direction. If a further change of direction is necessary, then:

1 provide the next size larger flue connector

2 increase the flue connector rise by 300 mm

3 deduct 10% for each additional change of direction from the listed flue capacity in the flue connector table.

Sizing the common flue

When sizing the common flue, the maximum common flue table is used.

Example

Four water heaters are to be installed on the ground floor of a four-storey building and connected through a manifold to a common flue (Fig 9.10).

Each appliance has a 100 mm flue outlet and an hourly gas consumption of 50 MJ and will operate independently.

The space available limits the flue connector rise to 600 mm. The spacing between the flue connectors is 750 mm.

Step 1 The flue connector size must first be determined from the flue connector pipe size table for low heat loss. The total height from the appliance draught diverter to the flue terminal is 18 m. In order to have a rise in the manifold, it is assumed that the connector rise of the appliance farthest from the common flue is 300 mm.

    From the maximum flue connector pipe size table for low heat loss with a total height of 18 m, a rise of 0.3 m, a 100 mm diameter flue connector has a capacity of 70 MJ, which is adequate.

Step 2 The manifold must be sized as a common flue since all appliances do not operate simultaneously. Using the maximum common flue size table for low heat loss the Type L line is used. For a total height of 18 m, on the L line, a 150 mm diameter flue has a capacity of 273 MJ, which is greater than the 200 MJ input of the appliances.

    A 125 mm diameter flue cannot be used as it has a capacity of only 188 MJ.

Step 3 Check that the manifold length A does not exceed 50% of total flue height.

FLUE INSTALLATIONS

The installation of the flue pipe is a very important factor to consider when selecting or recommending the appliance location. Consideration must be given to the difficulty in manipulating the flue pipe and the need to avoid changes of direction and lateral lengths. The following additional points must also be adhered to:

1 Locate the appliances to avoid cutting of joists and rafters.

2 Flues passing through combustible materials must have suitable clearances. Refer to the AS 5601 code and local authorities for acceptable clearances.

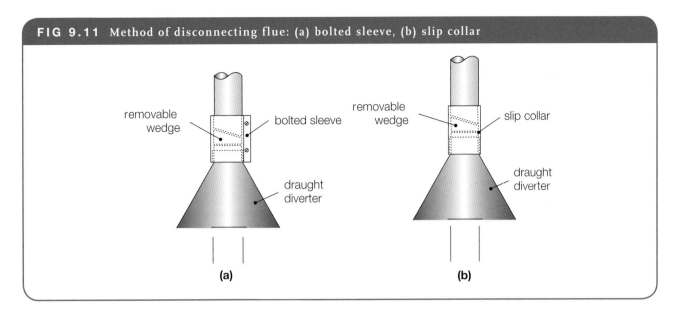

FIG 9.11 Method of disconnecting flue: (a) bolted sleeve, (b) slip collar

(a)

(b)

3 The flue pipe must be supported so that the weight of the flue is not on the appliance or the flashing.

4 A means of disconnecting the flue must be provided so that the appliance can be removed without disturbing the flue (Fig 9.11(a) and (b)).

5 Flue joints must be sealed with the socket facing upwards towards the terminal. The only exception is when a socket is used to slide over the top of the flashing, thus providing weathering at the top edge of the flashing.

6 Where necessary a condensation cone and drain must be provided—refer to the AS 5601 code or local authority. Condensation is best avoided by ensuring adequate draught diverter dilution and the avoidance of high heat loss situations.

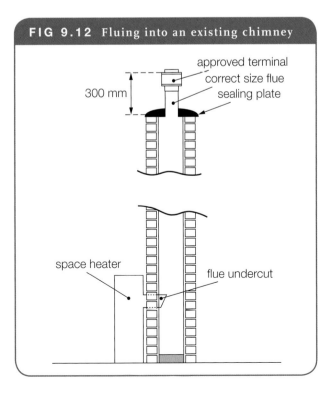

FIG 9.12 Fluing into an existing chimney

300 mm

approved terminal
correct size flue
sealing plate

space heater

flue undercut

7 Use of existing chimneys is acceptable providing that:

(a) They are checked for adequate size.

(b) Damper plates are removed or fixed into a full open position.

(c) The flue pipe does not extend into the chimney so as to cause a restriction or allow any loose materials which may fall down the chimney to enter the flue (Fig 9.12).The flue should be undercut a reasonable distance up from the bottom of the flue.

(d) The chimney must be clean, airtight and terminate above the roof line.

8 Power flues or flues terminating under a ventilated canopy should be specifically approved by the local authority before beginning the installation.

Flue terminal location

Correct location of the flue terminal is essential if products of combustion are to pass freely through the terminal and down draught conditions are to be avoided.

The following points should be considered when locating a flue terminal:

1 Terminate in a clear exposed position at least 500 mm from the nearest part of the roof.

2 Avoid locating terminal in roof gullies.

3 If the flue terminates on an outside wall below the roof line, the local authority must be consulted.

4 Flue terminals must not be vented near openings into buildings. The distances from openings are specified in the AS 5601 code and depend on the megajoule input rating of the appliance.

5 Termination in a roof space requires the approval of the local authority.

6 If the roof area is used by the occupants or public, the terminal must be at least 2 m above the roof and 500 mm above any wall or fencing surrounding the roof.

FIG 9.13 (a) and (b) Balanced flue system

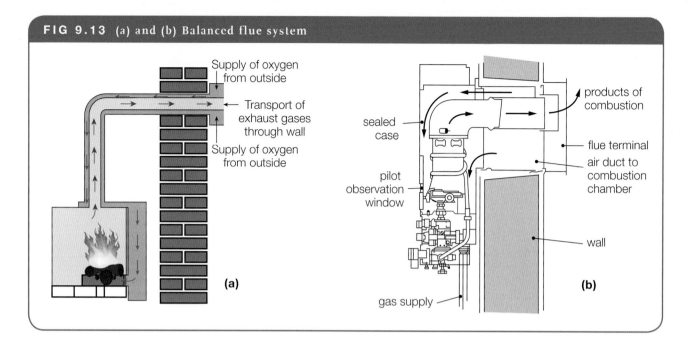

Supply of oxygen from outside

Transport of exhaust gases through wall

Supply of oxygen from outside

(a)

sealed case

pilot observation window

gas supply

products of combustion

flue terminal

air duct to combustion chamber

wall

(b)

BALANCED FLUED APPLIANCES

Principle of operation

Balanced flue systems are energy- and space-efficient and have become widely used for room and water heating. The balanced flued appliance draws its air for combustion from outside the building from a point immediately adjacent to where the products of combustion are discharged. The air inlet duct and outlet duct for the products of combustion are housed in the same terminal and are therefore exposed to the same atmospheric pressure. The appliance is sealed in the airtight compartment from the room. When no gas is being used, the pressure in the inlet and outlet ducts are equal and no air movement takes place. When combustion takes place, the hot flue gases become lighter and rise up the outlet duct and discharge to atmosphere; the colder heavier air surrounding the inlet duct replaces the hot flue gases, and thus a continuous supply of air for combustion is guaranteed (Fig 9.13).

Location of the balanced flue terminal

The correct location of the balanced flue terminal is very important. A bad location may cause the products of combustion to enter the building or, if too close to an obstruction such as eaves or return walls, the wind may cause the products to be blown down the air inlet duct and cause flame vitiation. The location of the balanced flue terminal shall comply with the requirements of AS 5601.

POWER FLUES

Power flues are classified as either forced draft or induced draft.

Forced draft flues have a fan that blows air into the base of the flue forcing the products of combustion through the flue to the terminal. Induced draft flues either have the fan at the terminal or in the lateral, or draw the products of combustion out of the flue.

As the flow of flue gases is reliant upon the operation of the fan it must be interlocked with the gas supply. In the event of a fan failure the gas supply is shut off.

VENTILATION REQUIREMENTS

Ventilation must be sufficient to guarantee adequate air for combustion and draught diverter dilution. The ventilation must also ensure a safe ambient temperature and the maintenance of a good environment in the room or area surrounding the appliance. If forced ventilation is used to provide air for combustion, draught diverter dilution and ventilation, automatic shutdown of the gas supply must take place in the event of a breakdown in the equipment providing the forced ventilation.

Normally, the local building regulations will ensure adequate ventilation for gas appliances, but particular care must be taken to ensure adequate ventilation when appliances are being installed in small rooms, or confined spaces such as cupboards, small laundries or boiler rooms. Ventilation requirements laid down by the AS 5601 code and the local authority must be adhered to.

Ventilation may be provided by adventitious opening, that is, normal spaces around doors and windows provided that the room volume is of adequate size. The maximum number of megajoules for all appliances installed in one room is not exceed 3 MJ for each cubic metre of room volume. For example, a room measuring 3 m × 3 m × 2.4 m would have a volume of 21.6 m³. If you multiply this by the maximum of 3 MJ per cubic metre, then the total capacity of the appliances installed in that room is 64.8 MJ. If the total input exceeds this amount, permanent natural ventilation is required.

The free space size of the permanent ventilation is calculated by multiplying the total MJ by a factor. The value of the factor is dependant upon the room type and whether the ventilation is drawn from the outside atmosphere or an adjacent room. Two vents are required—one at high level and one at low level. Each vent needs to be within 5 per cent of the room height measured from the top of the vent free space for the high level vent and the bottom of the vent free space for the low level vent.

For most rooms where the ventilation is direct to the outside atmosphere, you use the formula:

$A = 3 \times T$.

Where: A = free area of the vent in square centimeters

3 = factor when ventilation is drawn from the outside atmosphere

T = total megajoules rating of all the appliances

Example

A continuous flow water heater (150 MJ) is installed in a laundry measuring 3 m × 3 m × 2.4 m.

Room volume equals $3 \times 3 \times 2.4 = 21.6$ m³.

The total number of MJ supplied through adventitious openings would be $3 \times 21.6 = 64.8$ MJ.

As 150 is greater than 64.8, permanent ventilation is required. The size of each vent is calculated as follows:

$A = 3 \times T$

$A = 3 \times 150$

$A = 450$ cm²

Therefore each vent would have a free space area of 450 cm². When calculating the actual size of the vent an allowance must be made for obstructions such as grating or mesh.

If in the example above the ventilation was to be provided by an adjacent room, the vents would need to be bigger and are calculated using the formula: $A = 6 \times T$. Therefore, the vents above would be twice the size, that is, 900 cm². Refer to AS 5601 for further information.

FIG 9.14 Balanced flue terminal location

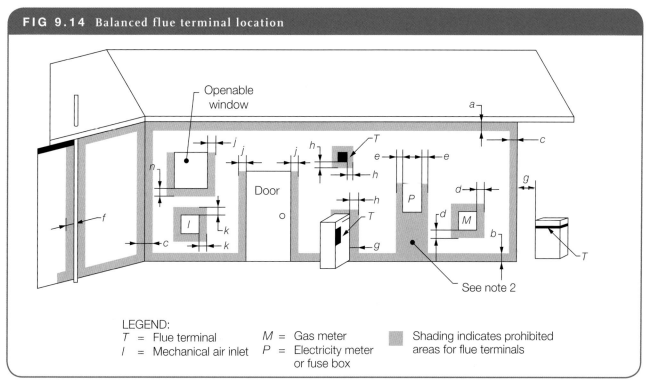

LEGEND:
T = Flue terminal
I = Mechanical air inlet
M = Gas meter
P = Electricity meter or fuse box
Shading indicates prohibited areas for flue terminals

Note:

1   Appliances with HGC above 150 MJ to terminate no less than 1500 mm in any direction from an openable window or air-intake.

2   In Victoria, appliances to terminate no less than 1500 mm horizontally and 2000 mm directly above or below a gas meter.

Minimum clearances required for balanced flue terminals, fan-assisted flue terminals, room-sealed appliance terminals or the terminals of outdoor appliances.			
		Minimum clearances (mm)	
Ref.	**Item**	**Natural draft**	**Fan assisted**
a	Below eaves, balconies and other projections:		
	• *Appliances* up to 50 MJ/h input	300	200
	• *Appliances* over 50 MJ/h input	500	300
b	From the ground, above a balcony or other surface †	300	300
c	From a return wall or external corner †	500	300

Ref.	Item	Minimum clearances (mm)	
		Natural draft	Fan assisted
d	From a gas meter (M) (see 4.7.11 for vent terminal location of *regulator*)	1000	1000
e	From an electricity *meter* or fuse box (P)	500	500
f	From a drainpipe or soil pipe	150	75
g	Horozontally from any guilding structure † or obstruction facing a terminal	500	500
h	From any other *flue terminal*, cowl, or combustion air intake †	500	300
j	Horizontally from an openable window, door, non-mechanical air inlet, or any other opening into a building with the exception of sub-floor ventilation:		
	• *Appliances* up to 150 MJ/h input	500	300
	• *Appliances* of 150 MJ/h input up to 200 MJ/h input	1500	300
	• *Appliances* over 200 MJ/h input up to 250 MJ/h input †	1500	500
	• *Appliances* over 250 MJ/h input †	1500	1500
	• Fan-assisted *flue appliances*, in the direction of discharge	–	1500
k	From a mechanical air inlet, inlcuding a spa blower	1500	1000
n	Vertically below an openable window, non-mechanical air inlet, or any other opening into a building with the exception of sub-floor ventilation:		
	• Space heaters up to 50 MJ/h input	150	150
	• Other *appliances* up to 50 MJ/h input	500	500
	• *Appliances* over 50 MJ/h input and up to 150 MJ/h input	1000	1000
	• *Appliances* over 150 MJ/h input	1500	1500

† Unless *appliance* is *certified* for closer installation
Notes:
1   All distances are measured to the nearest part of the terminal.
2   Prohibited area below electricity meter or fuse box extends to ground level.

CHAPTER TEN

Pipe installations

Pipe installations for gas must be carried out by an authorised installer. The basic standards for the design and the location of pipes and materials to be used for gas installations are governed by an Australian Standard. Administrative, gas supply and metering requirements are determined by the local authority. The Australian Gas Association (AGA), the Australian Liquefied Petroleum Gas Association (ALPGA) and Standards Australia have prepared an installation code for gas burning appliances and equipment. This provides essential requirements and basic standards for adoption by enforcing authorities. The authorised installer should purchase a copy of this code (AS 5601) before carrying out any gas work and confer with the local authority for any specific regulations enforced by that authority which are not covered by the code for gas burning appliances and equipment.

It is also important that the gasfitter obtain all relevant information prior to commencing work. Information can be obtained from relevant regulations, manufacturers, plans and specifications, as well as the client. Safety is of paramount importance when working with gas and all work should be carried out in a tradelike manner. When cleaning up the site observe environmental and regulatory recommendations and dispose of waste in an appropriate manner. It is important that the service is properly purged to ensure the safe operation of the system and to ensure that pipework is safe to work on when repairing, altering or extending pipework.

SCOPE OF THE AS 5601 CODE

This code provides basic standards for the design, fabrication, installation and operation of piping systems within the boundaries of the customer's property and the installation of appliances for use with fuel gases such as natural gas, LPG, SNG or TLP. It is not intended to cover systems or portions of systems supplying equipment engineered, designed or installed for specific manufacturing, production, processing or power-generating applications. It is not intended to apply to the internal construction of appliances which are provided for in other codes.

This code covers work from the outlet of the meter to the inlet of the appliance including flueing, ventilation, combustion air and the commissioning of the appliance. The code also covers requirements for gas installations in caravans and marine craft. The AS 5601–2004 Standards and/or local authority regulations covers work up to and including the meter for single meter installations in particular locations.

The requirements of the AS 5601 code are to be read in conjunction with, but do not take precedence over, any statutory regulation which may apply in any area. It is important that you are fully conversant with this code as it sets out the minimum requirements for the location, ventilation and installation of consumer piping (including purging and testing).

PURGING OF GAS SERVICES

Purging is the replacement within a vessel of one medium with another. For gasfitting this is either replacing air with gas when first turning the gas on or when refilling pipework OR replacing the gas with air or an inert gas to work on existing lines. Refer to AS 5601 for the method of purging small installations of less than 30 litres (0.03 m³) and the local authority for installations larger than 30 litres. The gasfitter must be aware at all times of potential ignition sources and where the purged gas collects as it could potentially create an explosive situation. Sufficient time must be allowed for the gas to dissipate before lighting any appliance. Remember that natural gas will rise and LPG will fall. If you can detect the presence of gas then assume that a combustible mix is present and provide ventilation to allow it to dissipate completely.

TESTING OF GAS SERVICES

It is important that every join on a gas service is tested for soundness prior to charging with gas and/or lighting an appliance. New pipework is pressure tested using air or an inert gas, such as nitrogen, to a minimum pressure for a minimum time limit. (Refer to the code for minimum pressure and times.) The minimum time of testing is dependant upon the volume of the installation. Pressurisation of the installation is achieved by means of a hand pump, compressor or compressed inert gas. Under no circumstances is an oxygen cylinder to be used to test an installation as it will create a combustible mixture within the pipework when filling the line with gas and could cause an explosion.

It is advisable to test an existing installation for soundness prior to commencing any working on it such as adding or replacing an appliance. This can be done by either conducting a drop test with a manometer or by using a bubble leak detector (see Fig 10.1) as it will check even the joins which are not exposed. Exposed existing joints could also be tested using soapy water, leak detection fluid or an electronic gas detector. This method of testing may also be used for pipework once the meter and appliances have been

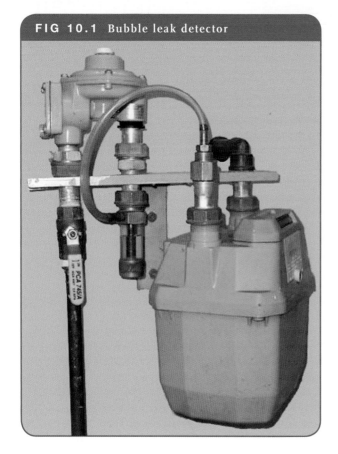

FIG 10.1 Bubble leak detector

connected. Refer to the code for a detailed explanation of testing procedures.

An example of the stages involved for testing a new installation.

1 Pipework (no meter or appliances connected)—pressure test with air or inert gas and Bourdon gauge.

2 Meter and appliances connected—bubble leak detector or drop test with manometer.

3 Testing of meter connections after the removal of bubble leak detector or reconnection—soapy water/leak detection fluid.

4 Checking of test point after checking of service or appliance regulator pressure—soapy water/leak detection fluid.

Electronic detection equipment may also be used but the operator needs to be fully conversant with this equipment before using it to detect gas leaks.

METHOD OF SIZING PIPE INSTALLATIONS

For appliances to operate efficiently, the installer must ensure that there is adequate volume and pressure of gas at the appliance at all times. To ensure that this is done economically, the minimum size pipe capable of supplying adequate gas and pressure should be used. The sizing of pipe installations can be done using a mathematical formula and calculations but this is difficult and time-consuming, so to simplify the process pipe-sizing charts are available in Appendix F of the AS 5601 code.

Factors to be considered when pipe sizing

Consideration must first be given to factors that affect the flow of gas through pipes.

1 *Length of pipe.* The pressure loss is directly proportional to the length of pipe. For example, if 10 m of a 15 mm steel pipe has a pressure loss of 25 Pa and if this 15 mm steel pipe was extended to 20 m, then the pressure loss would be 50 Pa.

2 *Number* of *fittings used.* Tees, elbows and bends offer more resistance to gas flow than a straight pipe. The equivalent effects of tees, elbows and bends are shown in Table 10.1.

3 *The specific gravity of the gas:* The heavier the gas the greater the resistance to flow. Therefore more pressure will be needed to push a gas such as propane with a *RD* of 1.5 through the pipe than natural gas with a *RD* of 0.6. A greater pressure is also needed at the appliance to push the heavier gas through the injector nipple at the burner.

4 *Material type of the pipe or tube:* Copper and plastics have a smoother bore than steel pipe and therefore offer less resistance to flow.

5 *Nominal size of pipe:* The size of steel pipe is measured by its internal diameter whereas the size of copper is measured by its outside diameter therefore steel pipe has a greater volume compared to copper of equivalent nominal size.

The following factors must also be taken into consideration when pipe sizing:

1 Consider future needs such as, is the customer likely to purchase an additional appliance in the future or replace an appliance with one of a much higher MJ rating? Making adequate provisions to meet that need now may save enlarging the existing system later.

TABLE 10.1 Equivalent effect of elbows and bends

Nominal pipe size (mm)	Equivalent length of pipe (m)	
	Elbows or tees	Bends
15	0.45	0.31
20	0.6	0.4
25	0.84	0.51
32	1.04	0.73
40	1.37	0.82
50	1.68	1.04
65	2.00	1.28
80	2.44	1.56
100	3.10	2.14

Note: The charts used in the AS 5601 code have an allowance for fittings built into them. No additional allowance is necessary under normal conditions.

2 In determining the size of piping by using the gas flow charts, the following information must be obtained:

(a) the type of gas, heating value and specific gravity

(b) the gas rate in megajoules per hour of each appliance to be used on the installation

(c) the pipe configuration, lengths of each section, number and types of fittings

(d) the static pressure of the installation

(e) meter outlet pressure and/or allowable pressure loss of the installation; different meter outlet pressures (i.e. pressure at the beginning of the consumer piping) have different allowable pressure drops to ensure adequate pressure on the inlet side of the appliance regulator. Refer to your local authority for the applicable meter outlet pressures and/or allowable pressure drops to be used

(f) diversity factors, arising from the probability that all appliances may not be in use at any one period of time. (Diversity factors are not normally allowed for in domestic homes but are usually allowed for in blocks of flats. The installer should consult the local authority with regard to allowances for diversity factors.)

3 When adding appliances to an existing installation, check to ensure that the meter, regulator and inlet service are capable of meeting the additional load. The pipe size of the existing installation also needs to be checked for its capability of carrying the additional load.

EXAMPLES OF SIZING PIPE INSTALLATIONS

Use of pipe-sizing charts

Note: Check with your local authority as to what meter outlets pressure and/or pressure drop to use for the relevant metering pressure.

Step 1 Sketch the pipe configuration including the length of all sections and the gas rates of each appliance in megajoules per hour (Fig 10.2).

Step 2 Determine the index length which is the distance along the main pipe run from the meter to the farthest appliance. This index length is the only length to be used when sizing all sections of the installation. If the index length falls between two lengths indicated in the tables, use the longer length.

Step 3 Commencing with the letter A at the meter, mark the index length into sections and then indicate the appliance connection point with a letter (Fig 10.3).

Step 4 Prepare a table with headings as shown:

Pipe section	Gas rate/hour (MJ/h)	Pipe size (mm)

Step 5 Select the appropriate pipe-sizing table for the nominated gas type, pipe material, supply pressure (meter outlet pressure) and/or allowable pressure drop.

Solution to Figure 10.2

Assuming the gas being used is natural gas with an HV of 38 and an RD of 0.6, size the pipe configuration shown in Figure 10.2. It is a low pressure system with an

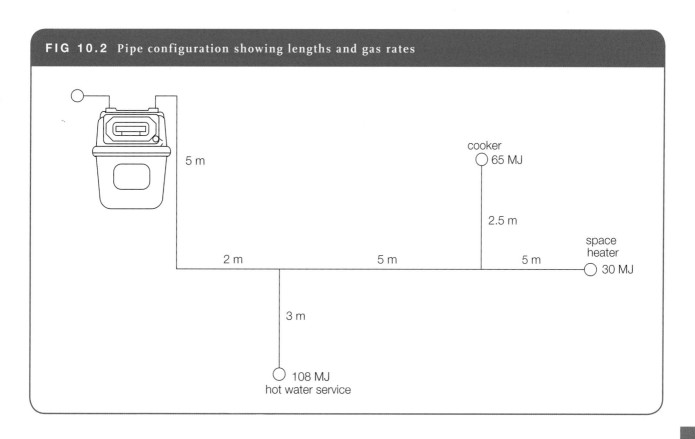

FIG 10.2 Pipe configuration showing lengths and gas rates

5 m

2 m   5 m

cooker
65 MJ

2.5 m

space
heater
5 m   30 MJ

3 m

108 MJ
hot water service

allowable drop of 0.25 kPa, which is suitable for meter outlet pressures (supply pressures) of between 1.5 and 2.5 kPa. Copper pipe is to be used. Using this information the table to be used is F3.

Calculate index length

The index length for Figure 10.2 is from point A to point D and measures 17 m (5 + 2 + 5 + 5). As 17 m falls between two lengths shown in Table F3 select 18 m, which is the longer length.

Size each section

Using the 18 m length and sizing section C-D, go down vertically under the 18 m section until a figure equal to the 30 MJ is found, or the nearest above it. In this case the figure of 39 MJ must be used. Carry along horizontally from the 39 MJ to the left-hand column and the size indicated is 15 mm. The correct size pipe for section C-D is 15 mm; mark this in the pipe-sizing chart. Follow the same procedure and size the other sections (see Fig 10.3).

Gas rates

Appliance	Gas rate (MJ/h)
Hot water service	108
Cooker	65
Space heater	30

Pipe-sizing chart (index length is 17 m)

Pipe section	Gas rate (MJ/h)	Pipe size (mm)
A-B	(95 + 108) 203	25
B-C	(65 + 30) 95	20
B-F	108	20
C-E	65	20
C-D	30	15

Hint: Commence with section farthest from meter and work back, starting with the individual branches then adding up MJ loading for the sections as you go.

Pipe sizing two stage installations

Pipe sizing two stage installations uses the same principles as single stage installations, except that each stage is sized using tables applicable to the pressure rating of each section.

The index length for the first stage section is from the regulator to the most disadvantaged second stage regulator and, conversely, the index length for the second stage is from the second stage regulator to the most disadvantaged appliance it serves.

As pressure setting of the first stage regulator is relatively high a much higher pressure drop is used, which results in smaller pipe sizes than lower pressure regulators. Refer to the tables as to which pressure drop is applicable.

Example

For the LPG installation in Fig 10.4, assuming the administration block had a gas loading of 200 MJ/hr and the classroom buildings a loading of 120 MJ/hr with an index length of 200 metres of copper piping, the pipe sizes would be indicated as follows.

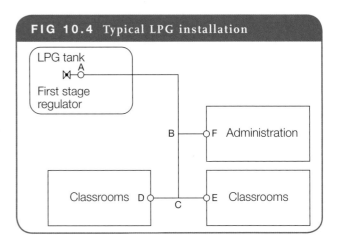

FIG 10.4 Typical LPG installation

If the first stage regulator is set at 70 kPa, use the table supplied in AS5601 for 10 kPa pressure loss.

Index length: 200 metres.

Section	MJ loading (MJ/hr)	Pipe size (mm)
A-B	(240 + 200) 440	20
B-C	(120 + 120) 240	20
B-F	200	20
C-D	120	15
C-E	120	15

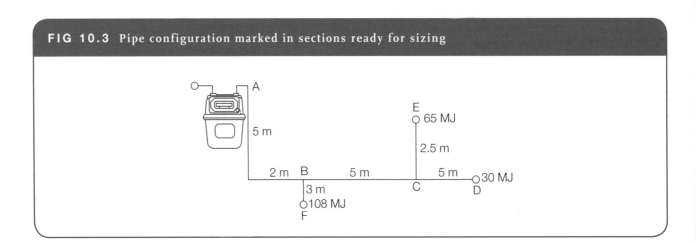

FIG 10.3 Pipe configuration marked in sections ready for sizing

If the first stage regulator was changed to 140 kPa, the pipe sizing would be as follows using the table supplied in AS 5601 and the table for 20 kPa pressure loss.

Index length: 200 metres.

Section	MJ loading (MJ/hr)	Pipe size (mm)
A–B	440	20
B–C	240	15
B–F	200	15
C–D	120	15
C–E	120	15

Use the same method to size the second stage pipework. The index length is measured from the second stage regulator to the furthest appliance.

Example

If the administration block had the layout shown below, use the LPG table for copper pipe. Supply pressure equals 3 kPa.

Appliance	Gas rate (MJ/h)
D—Hot water service	140
E—Space heater	30
F—Space heater	30

Index length: (5 + 10 + 2 + 4) 21 metres

Section	MJ loading (MJ/hr)	Pipe size (mm)
A–B	(170 + 30) 200	25
B–C	(140 + 30) 170	25
B–F	140	20
C–D	30	15
C–E	30	15

In this example the piping from the outlet of the second stage regulator ('A' in Figure 10.5) would be 25 mm. The piping leading into the higher pressure inlet would be 15 mm. That is, for the second stage regulator the inlet size at 140 kPa is 15 mm and the outlet size at 3 kPa is 25 mm. The second stage piping is usually bigger than the first stage due to the difference in pressures.

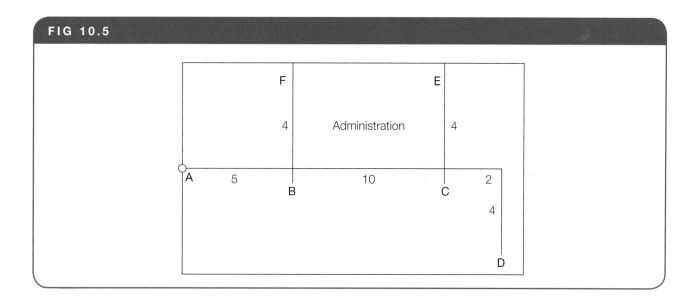

FIG 10.5

CHAPTER ELEVEN

Operation and commissioning of appliances

COMMISSIONING OF APPLIANCES

Appliances are commissioned to ensure that they are performing safely and to their design requirements. In general terms commissioning involves the following:

1 Check that the installation complies with the manufacturer's instruction and applicable gas standards and regulations.

2 Test the installation for soundness (bubble leak test, drop test and/or soapy water/leak detection fluid).

3 Purge the gas line to ensure that all air is removed prior to lighting up. This is achieved either through an open burner or by loosening the connection union and waiting until there is an audible change in sound and/or the smell of gas. Retest union after tightening with soapy water/leak detection fluid. *Note:* **Wait for the gas to dissipate before lighting the appliance (this may take several minutes).**

4 Check that all the burners and accessories are in their correct position.

5 Check appliance pressure with at least 50% of the burners alight. For natural gas appliances attach a manometer to the test point on the appliance regulator or an injector nipple. On LPG appliances attach the manometer to the test point at the cylinder regulator, test point supplied with appliance or injector nipple.

6 Check the gas consumption of the appliance against its rated performance (this can only be done on installations that have a gas meter). See GR formula on p. 74.

7 Check operation of thermostats, safety devices and all accessories such as timers.

8 Instruct the customer on the operation of the appliance and leave the instructions with them.

It is advisable to have a record sheet to record the results of the commissioning process. If an appliance is found to be outside of the 10% performance range (check the data plate for consumption) firstly re-do the calculation over a longer period, if still incorrect inform the client and manufacturer. If an appliance pressure is too high it can burn out the combustion chamber, shortening the life of the appliance, or even cause incomplete combustion due to over-gassing. If it is too low then the appliance will not provide the heat required and give poor performance.

The following is an example of appliances commonly installed and commissioned but not an exhaustive list.

COOKERS

Types of cookers

The function of a cooker is to cook food by one of the following methods:

1 boiling or frying on the hotplate

2 grilling

3 baking in the oven.

The main types of cookers are as follows.

1 Upright. This combines all three methods of cooking in one appliance by placing the oven, grill and hotplate on top of one another. It has the advantage of providing all three methods of cooking while taking up the minimum of floor space (Fig 11.1).

FIG 11.1 Upright cooker

FIG 11.2 Range

FIG 11.3 Elevated cooker

FIG 11.4 Wall oven and grill

2 Range. This is a wider version of the upright, often with the convenience of a second oven and griddle plate (Fig 11.2).

3 Elevated cooker. The oven is placed by the side of the hotplate and grill instead of underneath. A cabinet must be provided to support the cooker at the correct height. Elevated cookers can be purchased with the oven to the right or left of the hotplate and grill section. The main advantage of this type of cooker is that the oven is situated at a more convenient height (Fig 11.3).

4 Wall oven and grill. Consists of an oven and grill compartment which are either fitted into a cabinet or built into a wall (Fig 11.4).

5 Built-in hotplate. Consists of a hotplate which is fitted into the work top of a cabinet.

PRINCIPLE OF OPERATION

Hotplate

The hotplate consists of an aerated burner and a means of supporting the cooking utensil above the burner (trivet). Heat is transferred from the flame to the utensil by radiation, convection and conduction. Radiation accounts for a very small percentage of the heat transferred to the utensil. Convection in the form of the hot burnt gases leaving the intermediate zone of the flame is the main factor in the transfer of heat to the bottom and sides of the utensil. Conduction is the transfer of heat through the utensil to the food.

The trivet supports the utensil above the flame to ensure maximum efficiency in transferring the heat to the utensil. Flame contact should be made after the gases have completely burnt in the intermediate zone. The trivet must be robust and capable of supporting small and large utensils but not too bulky or it will absorb too much energy. Cookers usually have deep spillage bowls which fit tightly around the burner; secondary air for combustion is drawn from the space between the top of the trivet and the bottom of the spillage bowl (Fig 11.5).

FIG 11.5 Typical natural gas hotplate burner, spillage bowl and trivet

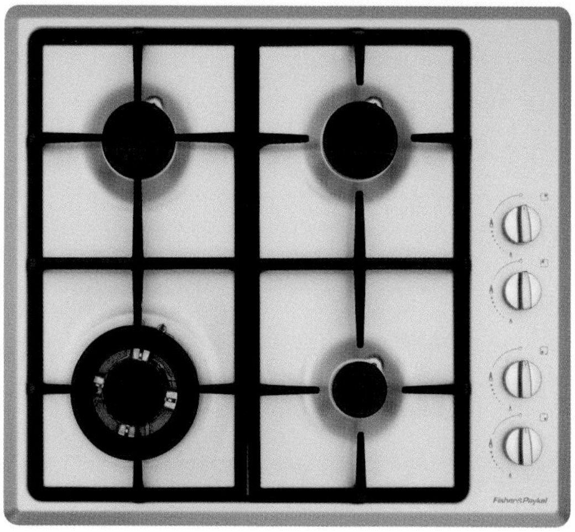

Under no circumstances must the inner cone of the flame come into contact with the utensil as this will cause incomplete combustion to take place. The trivet is designed to keep the utensil well clear of the inner cone.

Grill

Food cooked with a griller is heated from above with radiant heat (Fig 11.6).

The surface combustion burner used by grills has the advantage of having an even heat over the whole grilling surface and consists of an aerated burner with an enlarged burner head which forms the complete area of the grilling surface. The burner head is made from either a perforated enamelled sheet steel or a fine steel mesh. Gas passes through the perforated surface with combustion taking place on the surface. The perforated enamel sheet steel type does not glow a bright red but the food is cooked by direct flame radiation and high flame temperature. The fine steel mesh burners glow a bright red which radiates heat onto the food. Surface combustion grillers use a very high percentage of primary air. If burnt with insufficient primary air the flame tends to float over the surface and lap around the side of the burner.

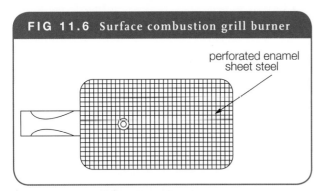

FIG 11.6 Surface combustion grill burner

perforated enamel sheet steel

Oven

The oven consists of a box and a burner with provisions made to draw in air for combustion and an outlet for discharge of the products of combustion. The food is cooked by convection currents created by the circulation of the products of combustion and by radiant heat from the sides and top of the oven.

Development of the oven

There are five main stages of development.

Stage 1 Refer to Figure 11.7.

This early type of oven had no insulation or thermostat. The convection currents from the hot gas had very little work to do before being vented at the top of the oven, making the oven very inefficient. The food cooked very unevenly with a tendency to overcook on the bottom. To counteract this, a metal plate was sometimes placed above the food and when the metal plate got hot, it radiated heat onto the top of the food.

Stage 2 Refer to Figure 11.8.

A slight improvement in design. Some of the ovens were insulated and thermostats were starting to be used. The

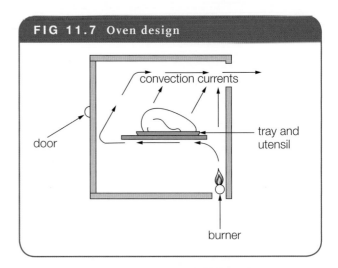

FIG 11.7 Oven design

convection currents

door

tray and utensil

burner

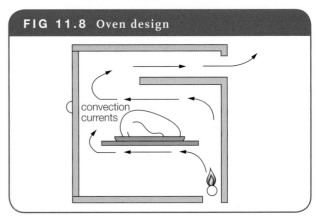

FIG 11.8 Oven design

convection currents

convection currents from the hot gases had a little more work to do which improved efficiency slightly. Although this design improved the cooking performance, the metal plate was still placed above to radiate heat onto the top of the food.

Stage 3 Refer to Figure 11.9.

These ovens were well insulated and thermostatically controlled. The hot gases had to circulate, giving off much of their heat before passing out of the oven. Because hot gases rise and then fall when cooling, the top of the oven was approximately 25°C hotter than the bottom. This type of oven was called a zoned heat oven and allowed several dishes to be cooked at slightly different temperatures at the

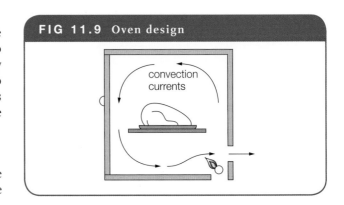

FIG 11.9 Oven design

convection currents

same time. It also made it necessary for the cook to locate the food on the correct shelf. The cooking results from this type of oven are excellent.

Stage 4 Refer to Figure 11.10.

This design is used on the majority of modern cookers. It functions in the same way as that described in Figure 11.9 except that it carries the products and food vapour away from the wall immediately behind the cooker and vents them into the room from the top of the cooker. All modern ovens have a safety device to ensure the gas supply is shut off in the event of flame failure. The modern oven is available as a manual oven or it can have the added convenience of being fully programmed.

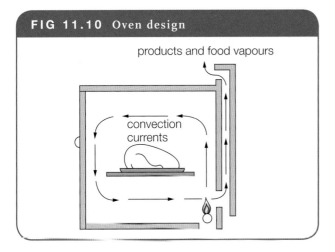

FIG 11.10 Oven design

products and food vapours

convection currents

Stage 5 Refer to Figure 11.11.

The convection oven uses a fan to give an even distribution of heat throughout the oven. Food can be placed on any shelf in the oven. Some convection ovens can be used as ordinary zoned heat ovens in the event of an electrical breakdown. Convection ovens which can only be used with the fan generally use the fan to circulate the hot gases and increase the pressure within the oven, thus reducing the cooking time by up to one-third. For example, a cooking time of 60 minutes in a normal cooker would be reduced to 40 minutes in this type of convection oven.

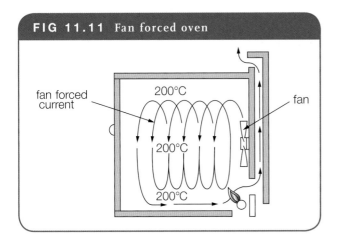

FIG 11.11 Fan forced oven

fan forced current

200°C

fan

200°C

200°C

OVEN VITIATION

Oven vitiation (flame suffocation) is caused through inadequate secondary air being available from around the burner, usually due to the products of combustion not passing, or only partially passing, up the flue, or being recirculated around the burner.

Causes of oven vitiation

1   The oven flue is blocked.

2   The gas rate is too high.

3   The angle of the burner is wrong or the burner is wrongly adjusted. If the burner is not fitted and adjusted correctly, the products of combustion may be projected forward onto the bottom of a utensil or the solid section of the oven shelf which some ovens have. This will cause some of the products to recirculate around the burner and thus exclude secondary air.

4   An oversized cooking utensil is used. The cooking utensil must have adequate space around it to allow the convection currents to circulate correctly or they will recirculate around the burner and exclude secondary air. This problem can also be identified by the food being overcooked on the bottom.

Self-clean oven

A catalyst material is mixed with the porcelain before it is sprayed onto the oven linings. The catalyst decomposes grease and cooking vapours during the cooking cycle. The decomposed matter changes to a fine ash. The catalytic effect takes place between temperatures of 150–350°C, and between these temperatures a continuous cleaning process takes place. Problems with grease and food vapour build-up will occur if the cooker is used often at low temperatures or excessive spillage of fats takes place. Because the bottom of the oven is the coolest spot, some manufacturers provide a tray to catch excessive spillage. Self-clean oven linings must only be wiped with a soft rag. Under no circumstances must abrasives or cleaning agents be used. If a build-up of grease occurs, set the oven to a high temperature and leave on until clean.

Rotisseries

The rotisserie is an additional feature of many ovens. It provides a means of barbecuing or broiling the food. This is done by passing a spit through the meat, securing one end to a support and locating the other end into a motor so that the meat can be rotated slowly while cooking. Other accessories include timers, clocks and various electronic ignition devices.

COMMISSIONING COOKERS

Installation check

1   Check that appliance installation complies with the manufacturer's instructions, the AS 5601 code and any special local authority's requirements.

2 Check that cooker is level and firm.

3 Check that wall ovens are secured to the cabinet.

Light up and check

1 Attach manometer to a suitable point on the outlet side of the appliance regulator.

2 Check badge plate for correct working pressure of appliance, and set pressure correctly with at least 50% of the burners full on.

3 Check and adjust the oven burner, and leave oven set at 150°C or equivalent.

4 Adjust hotplate burners and grill.

5 Check operation of ignition device/s.

6 When oven is hot, turn thermostat setting back to 120°C—burner should be at minimum bypass rate.

7 Carry out oven door slam test to ensure normal opening and closing of the door will not cause flame outage.

8 Check appliance performance by checking whether the actual gas consumption is within 10% of the data plate rating (all burners must be on high).

9 Check flame failure device is operating in the advent of flame failure.

10 Check operation of rotisserie, programmers and lights.

Instructions for customer

1 Instruct customer on correct lighting and operating procedures.

2 Be sure customers understand by getting them to demonstrate the lighting and operating procedures.

3 Draw attention to the manufacturer's written instructions.

INSTANTANEOUS HOT WATER SERVICES

Instantaneous hot water services do not store water; the water is heated as it flows through the hot water service. The capacity of the instantaneous hot water service is measured in litres per minute for water flow and degree centigrade for the temperature rise. An instantaneous hot water service is very compact and is available as an inside wall model with a conventional flue, or as an outside wall model using an inbuilt balanced flue system.

PRINCIPLE OF OPERATION

When a hot tap is turned on, the flow of water through the water section of the heater lifts the gas valve, allowing the gas to the burner to heat the water. When the hot tap is turned off, the gas also goes off. As the water flow increases the temperature of the water decreases and conversely as the water flow decreases the temperature increases as the burner flame size and resultant energy is constant.

Operation of the instantaneous hot water service

1 When no water is being used, the pressure on the bottom and top side of the diaphragm is equal.

2 When water is flowing through the water section, it increases its velocity when passing through the venturi.

3 The velocity of the water passing through the venturi sucks the water out from the top side of the diaphragm, thus creating a differential pressure between the top and bottom side of the diaphragm.

4 The higher pressure on the bottom side of the diaphragm causes the diaphragm, bearing plate and water spindle to rise, and the water spindle pushes the gas valve up, allowing gas to flow to the burner.

5 When the hot tap is turned off, water flows back into the top section of the diaphragm and this equalises the pressure between the top and bottom sides of the diaphragm.

6 A spring, which is usually placed on top of the gas valve, reacts to shut the gas off.

Two-stage ignition

It is essential that the gas lights in two stages. Stage one allows sufficient gas to pass to the burner to establish

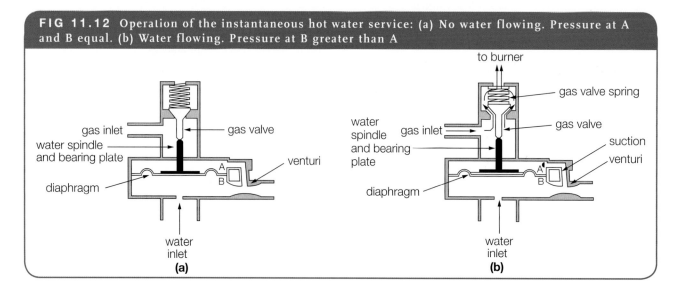

FIG 11.12 Operation of the instantaneous hot water service: (a) No water flowing. Pressure at A and B equal. (b) Water flowing. Pressure at B greater than A

a small stable flame on the whole of the burner. Stage two allows the gas rate to increase gradually to the full gas rate. If two-stage ignition is not achieved and the full gas rate is allowed to the burner in one lift, flash ignition will take place. This type of ignition will overheat the combustion chamber and heat exchanger. If flash ignition is not corrected, the working life of the combustion chamber and heat exchanger will be considerably shortened.

Operation of the slow ignition device (SID)

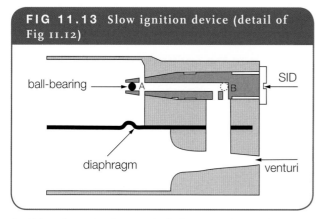

FIG 11.13 Slow ignition device (detail of Fig 11.12)

ball-bearing — A • □ B — SID

diaphragm — venturi

1   When the water is turned off, the water that pushes up into the section on top of the diaphragm also pushes the ball bearing to position A on the SID.

2   When the water is being used and therefore sucked out of the top section, sufficient water will pass through the two openings behind the ball bearing, allowing water to be drawn out of the top section to give a slight lift of the diaphragm. This slight movement of the diaphragm will be sufficient for the first stage of ignition.

3   As the ball bearing is sucked along the tube, it will shut off the withdrawal of water from the section on top of the diaphragm until it reaches position B.

4   When it reaches position B, the rest of the water will be sucked through the small opening allowing a full lift of the diaphragm and gas valve.

   Note: The two-stage ignition described in points 1 to 4 will take approximately 2 seconds.

5   When the water is turned off, the flow of water back into the section on top of the diaphragm quickly pushes the ball bearing to position A.

6   When the ball is in position A, water rushes into the top section, restoring equilibrium above and below the diaphragm, which allows the gas to shut off quickly. *Note:* If the gas does not go off quickly when the water is turned off, the small amount of water in the heat exchanger will increase in temperature rapidly. Therefore, it is essential that this is checked and corrected if the gas is not going off quickly.

Operation of the temperature selector

The temperature selector increases or decreases the volume of water flowing through the heater without affecting the gas rate.

Operation of the water governor

Once the temperature selector has been set, the water governor will ensure a constant water flow and temperature under varying inlet water pressures.

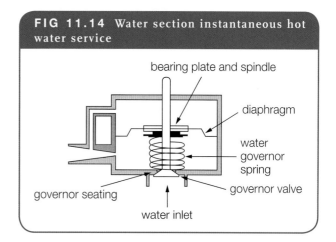

FIG 11.14 Water section instantaneous hot water service

bearing plate and spindle

diaphragm

water governor spring

governor seating

governor valve

water inlet

1   The water governor does not operate until the diaphragm has lifted and full gas is established at the burner. This is because the valve is so far away from the seating that it does not create sufficient resistance to water flow until the diaphragm has lifted.

2   The water governor spring ensures the governor valve follows the movement of the diaphragm.

3   When the diaphragm lifts, the position of the governor valve in relation to the valve seating will determine water flow.

4   Any increase or decrease in water pressure will cause the diaphragm and the water governor valve to move and therefore maintain a constant flow of water.

Additional control of ignition gas

To ensure the correct amount of gas is supplied for the first stage ignition, the gas is passed through a cross ignition bolt containing a fixed orifice.

When the diaphragm lifts and pushes up the gas valve stem, the stem passes through the small bottom valve but lifts the top valve. Gas must pass through the cross ignition bolt to get to the burner. This ensures the correct volume of gas is allowed to the burner for first stage ignition. The circlip on the gas valve stem pushes the bottom valve off its seating, allowing full gas to the burner when the diaphragm lifts fully in the second stage ignition cycle.

Installation check

1   Check that appliance installation complies with the manufacturer's instructions, the AS 5601 code and any special local gas or water authority requirements.

2   Check all hot water draw-off points for water flow.

Light up and check

1   Attach manometer, check and set working pressure with main burner full on.

2   With main burner off, check pilot flame size and position. This is to ensure it will light the gas on first stage ignition and heat up the flame failure device.

3   Turn on the check for two-stage ignition.

4   Observe main burner shuts off quickly when flow of water is stopped.

5   Turn on and check temperature of water at all outlet points.

6   After heater has been on for at least 5 minutes, check operation of flue.

Instructions for customer

1   Instruct customer on correct lighting and operating procedures.

2   Be certain the customers understand—have them demonstrate the lighting and operating procedures to you.

3   Leave the manufacturer's operating instructions with the customer.

CONTINUOUS FLOW HOT WATER SERVICES

Continuous flow hot water services do not store water; the water is heated as it flows through the hot water service. The capacity of the continuous flow hot water service is measured in litres per minute for water flow and the temperature can be adjusted to suit the application. With continuous flow hot water heaters the temperature of the water remains constant even if the water flow changes as its computer panel increases or decreases the flame size to match the water flow and thus maintain a constant temperature.

PRINCIPLE OF OPERATION

As water passes through the heater it is heated by the burner to a predetermined temperature. The gas flow and water flow are matched by the control panel in the heater so the temperature of the water is not affected by variances in flow.

Operation of the continuous flow hot water service

The flow of water through the water section of the heater activates a switch which in turn opens the gas valve and allows gas to enter the burner. The burner is then lit by an electronic igniter. Prior to ignition though the fan purges the combustion chamber of any residual gas. The control panel within the heater senses the rate of water flow and matches the flame size to this rate to deliver water at the pre-set or selected temperature. If the water flow changes e.g. another tap is turned on, the heater will adjust the

FIG 11.15 (a) Continuous flow hot water service (inside) and (b) outside

(a)

(b)

flame to accommodate for this. Temperature control panels can be added to the system so the temperature of the water can be selected by the user remotely at the source of usage. Where more than one temperature control panel is fitted the one fitted in the bathroom and/or ensuite will be dominant.

Operation of the temperature control

The temperature panel (where optioned and installed) allows the user to select the temperature of the water up to 50°C. The control panel sends a signal to the computer panel in the heater indicating the temperature required. Where not fitted the unit is usually pre-set to 50°C.

Installation check

1 Check that appliance installation complies with the manufacturer's instructions, the AS 5601 code and any special local gas or water authority requirements. (In particular check that the gas supply is adequate for the rating of the appliance).

2 Check all hot water draw-off points for water flow.

3 Check flue terminal distance from openings. (This distance can be a significant amount where the appliance rating is high (i.e. over 150 MJ/h)).

Light up and check

1 Attach manometer, check and set working pressure with main burner full on.

2 Check for efficient ignition.

4 Observe main burner shuts off quickly when flow of water is stopped.

5 Check temperature of water at all outlet points.

6 After heater has been on for at least 5 minutes, check operation of flue.

Instructions for customer

1 Instruct customer on correct lighting and operating procedures.

2 Be certain the customers understand by having them demonstrate the lighting and operating procedures to you.

3 Leave the manufacturer's operating instructions with the customer.

STORAGE HOT WATER SERVICES

The storage hot water service consists of an insulated cylinder which stores the water at a preset temperature ready for use. Storage heaters are thermostatically controlled and when hot water is drawn off from the top of the cylinder, cold water replaces it at the bottom. The incoming cold water enters close to the thermostat sensor so the thermostat will react and bring on the gas to replace the hot water drawn off.

PRINCIPLE OF OPERATION

Water becomes less dense when heated and therefore rises and is replaced by cooler denser water from within the heater. This circulation (hot rises, cold falls) is referred to as convection currents and results in the hottest water settling at the top and the cooler water at the bottom. This is known as 'stratification'. The thermostat is usually located close to the bottom of the cylinder and the cold water inlet. When the water around the thermostat has reached the set temperature, the gas is switched off and when the water around the thermostat has cooled sufficiently, full gas is restored to reheat. Because of this, an automatic permanent source of ignition is required to reignite the burner. The temperature differential between gas off at present temperature and gas back on for reheating is between 5 and 11°C, e.g. if gas is off at 55°C then it will come back on between 44 and 50°C.

Stacking

If all the water is drawn off and the total capacity of the unit is heated from scratch, the temperature differential between the top and bottom of the cylinder is small, often no more than 3°C. If small amounts are drawn off, the incoming cold water will cool the thermostat sensor allowing the gas to come on to reheat the replacement water. The thermostat may over-react and replace more energy than was drawn off.

If the heater is subjected to continuous small draw-offs, a considerable temperature differential between the top and bottom of the cylinder may occur. This is known as 'stacking'. To minimise stacking, some manufacturers have inserted a plastic dissipater into the cold water inlet. This spreads the cold water entering the unit and minimises the over-reaction of the thermostat.

Heat transfer

Refer to Figures 11.16(a), (b) and (c).

Figure 11.16(a) shows the traditional method of transferring heat from the hot gases leaving the flame into the water. Although it is an efficient method of transferring heat into the water, it also accounts for most of the heat losses when the water is up to temperature and the gas is off.

This method shown in Figure 11.16(b) creates a high turbulence inside the central flue, causing the products of combustion to recirculate repeatedly, so that more heat is transferred to the water. Heat losses are also less than those of heaters using the standard baffle.

There is no central flue in the method shown in Figure 11.16(c); the products of combustion transfer their heat into the water through the base and sides of the unit. The products have to push up the side of the cylinder and then down, before they escape through the balanced flue terminal. Because this takes longer and they are in contact with a larger area, more heat is transferred into the water. When the water is up to temperature and the burner is off, heat losses are minimised, because a trap is formed as the hot air from the cylinder has to push down against the cold air which forms in the down section of the heat exchanger.

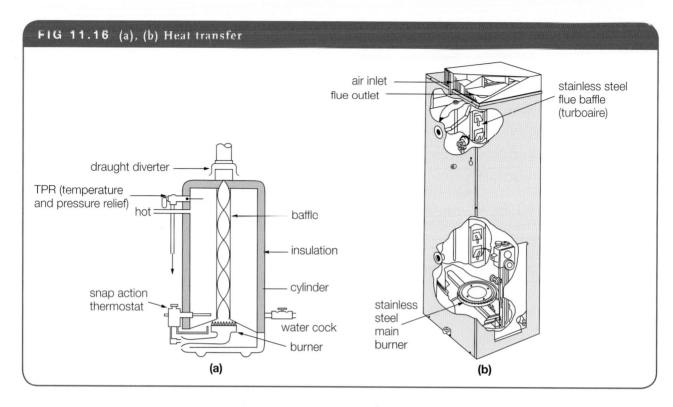

FIG 11.16 (a), (b) Heat transfer

(a)

- draught diverter
- TPR (temperature and pressure relief)
- hot
- snap action thermostat
- baffle
- insulation
- cylinder
- water cock
- burner

(b)

- air inlet
- flue outlet
- stainless steel flue baffle (turboaire)
- stainless steel main burner

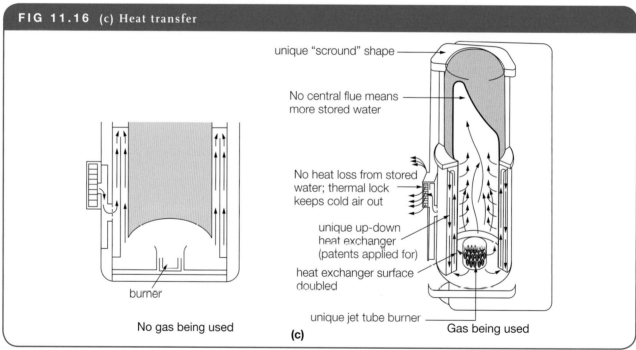

FIG 11.16 (c) Heat transfer

- unique "scround" shape
- No central flue means more stored water
- No heat loss from stored water; thermal lock keeps cold air out
- unique up-down heat exchanger (patents applied for)
- heat exchanger surface doubled
- burner
- unique jet tube burner

No gas being used   Gas being used

(c)

Capacity and recovery rates

It is important with gas storage heaters that both the capacity of the cylinder and the number of litres which can be reheated per hour is taken into consideration when sizing the unit. Slow recovery units usually have a large storage capacity to compensate for the slow recovery rate. Domestic hot water services with a fast recovery usually store a minimum of 45 L and a maximum of 135 L with a recovery rate of 90–145 L/ h.

TYPES OF HOT WATER SERVICES

Mains pressure fast recovery

This type of heater is usually connected direct to the domestic cold water service and is constructed of steel with a vitreous enamel coating which is referred to as a 'glass-lined' cylinder (Fig 11.17(a)).

FIG 11.17 (a) Outside fast recovery hot water service (foreground). (b) Inside fast recovery hot water heater with fitting for external flue (top), piezo igniter (bottom) and TPR (temperature and pressure relief)

Glass-lined cylinders

The vitreous enamelled surface is slowly soluble in water, and the solubility increases with the temperature of the stored water. To protect the steel from any imperfections in the coating, a sacrificial anode is used, which is colour coded for the specific water conditions of the geographical area it is to be installed in. It is important that the correct one for the area is used. These units must have a pressure and temperature relief valve fitted in the top 150 mm of water or in the top 20% of water capacity of the heater, whichever is the higher. The pressure and temperature relief valve can be combined and will relieve any pressure in excess of its set capacity, whether it is excess cold water service inlet pressure or thermal expansion. When the heater is in its heating cycle, thermal expansion will cause a steady drip of water from the pressure relief section of the valve. This will reseal when the heating cycle stops. Thermal relief takes place if the temperature of the water reaches 99°C. The water around the relief valve will have to drop considerably before it will reseal. In areas where the water supply contains a high calcium content, a cold water pressure relief valve should be fitted on the cold water inlet between the cylinder and the check valve. The fitting of a cold water pressure relief valve is compulsory in some areas.

COMMISSIONING STORAGE HOT WATER SERVICES

Installation check

1 Check that appliance installation complies with the manufacturer's instructions, the AS 5601 code and any local gas or water authorities' codes.

2 Check all hot water draw-off points for water flow.

Light up and check

1 Attach manometer, and check and set working pressure with main burner full on.

2 Check and adjust main burner aeration.

3 With main burner off, check and adjust pilot.

4 Check operation of thermostat and set to suitable temperature.

5 Check the appliance consumption against the data plate.

6 After heater has been on for at least 5 minutes, check operation of flue.

7 Check operation of pressure and temperature relief valve.

8 Check and time the operation of the safety device.

Instructions for customer

1 Instruct customer on correct lighting and operating instructions.

2 Be sure the customers understand by getting them to demonstrate the lighting and operating instructions to you.

ROOM HEATERS

The function of a room heater is to heat a room or area to a comfortable temperature and maintain comfortable conditions within that room or area.

Comfort conditions in a room

Comfort in a room depends upon the following factors:

1 Temperature. The temperature required for comfort in a lounge room varies from person to person, but a temperature of 22–23°C is generally accepted as being comfortable in Australia. For people to be comfortable, they must lose a small amount of heat. The temperature of exposed clothing and skin is usually 24–25°C; a room temperature of 1–2°C below this temperature will ensure people lose a small amount of heat to the room. There should not be excessive variation in temperature between floor and ceiling.

2 Ventilation. Sufficient ventilation to ensure an adequate supply of fresh air to the room is essential. Air movement without creating draughts is also necessary if the room is to be ideally comfortable.

3 Humidity. The relative humidity should not exceed 70%.

Method of heating a room

When heating a room or space either radiant or convection heat is used.

Radiant heat

Radiant heat travels in straight lines and does not give off its heat until it strikes a solid object (e.g. people, walls or furniture).

The air between the room heater and the solid objects is not heated by radiant heat, but when walls and furniture become warm, the air that comes into contact with them is heated, which gradually warms the air in the room (Fig 11.18(a)).

Advantages of radiant heat

1 It is not affected by draughts.

2 It can be directed where it is wanted.

3 It warms the furniture.

4 It has a warm red glow which has a psychological effect.

Disadvantages of radiant heat

1 Temperature is uneven in the room.

2 It is slow to bring the air temperature up to an even, comfortable level.

Convection heat

Air when it is heated becomes lighter; the lighter heated air rises and is replaced by cooler air. This movement sets up a continuous circulating cycle. A more efficient circulation of these air currents can be achieved by using a fan (Fig 11.18(b)).

Advantages of convection heat

1 It heats the room quickly and evenly.

2 There is continuous air movement.

Disadvantage of convection heat

It is affected by draughts.

TYPES OF ROOM HEATERS

Radiant forced convection heaters

This is a very popular type of heater which has all the advantages of both radiant and convection heat to ensure maximum comfort conditions in the space being heated. This style of heater is available with swank surface combustion radiant blocks, which ensure a high temperature radiant surface at all times. These surface combustion heaters are not thermostatically controlled but have available a manual control which allows a quick heating up cycle and a variety of turn down rates to ensure a comfortable temperature is maintained. The alternative is the type of heater that

FIG 11.18 (a) Radiant heat, (b) convection heat, (c) combination of radiant and convection heat

(a)   (b)   (c)

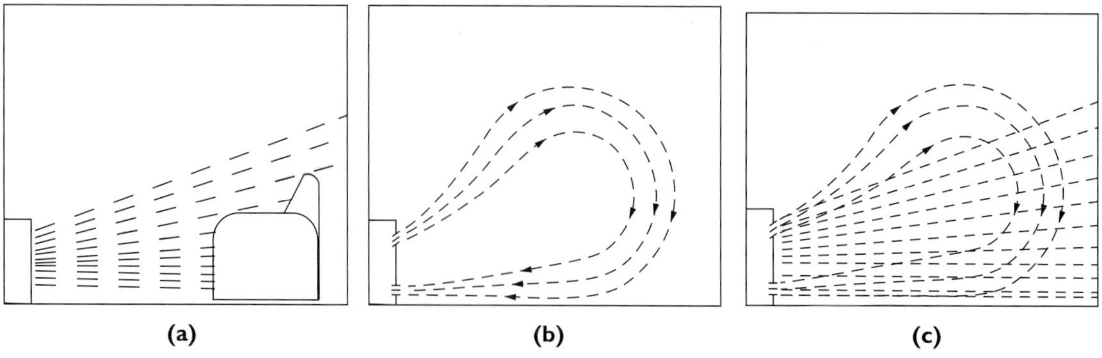

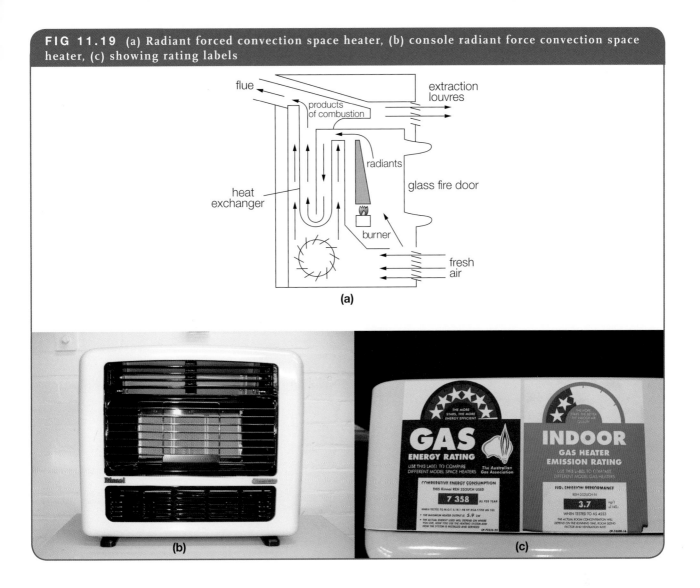

FIG 11.19 (a) Radiant forced convection space heater, (b) console radiant force convection space heater, (c) showing rating labels

uses fire clay radiants which are heated to a high temperature for radiant heat. These units are normally thermostatically controlled and when the room reaches the required temperature, the flame is reduced to the rate needed to maintain the temperature of the room. Radiant forced convection heaters are available as console models or as inbuilt models (Fig 11.19(a), (b) and (c)).

Radiant natural convection heaters (flued type)

Radiant convector room heaters are radiant heaters which have the fire frames designed with air passages to permit air from the room to be drawn in through openings at the base of the fire frame. This air is passed over a heat exchanger and discharged back into the room. The convected air does not come into contact with the products of combustion. These heaters are not quite as efficient as fan-forced convection heaters but are ideal for heating small rooms (Fig 11.20(a) and (b)).

Flueless heaters (domestic type)

These room heaters are normally a combination of radiant and convection heat. The flame and the products of combustion heat the radiant block and then the products of combustion are mixed with comparatively large volumes of air drawn in at the base of the heater, which are then discharged into the room through a grille at the top of the heater. Because of the amount of dilution that takes place, the products of combustion return to the room at a much lower temperature. Flueless heaters designed and approved for use with LPG incorporate an oxygen depletion device, which shuts off the supply of gas to the burner if the oxygen content of the room is affected by the products of combustion. Flueless heaters for installation in domestic homes using natural gas are not approved by all local authorities. When heaters have been approved for installation, the AS 5601 ventilation code must be strictly adhered to.

Forced convection heaters (flued type)

Forced convection heaters use a fan to draw air into the bottom of the unit. This air is then passed over a heat exchanger and then pushed out the top grill section.

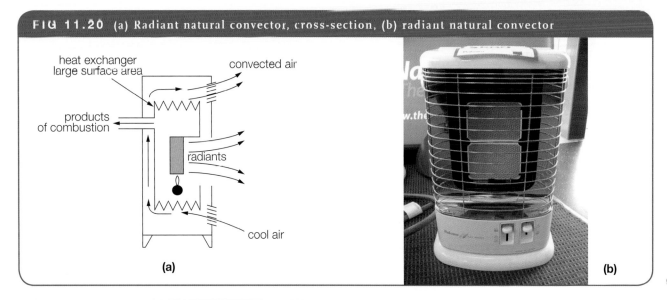

FIG 11.20 (a) Radiant natural convector, cross-section, (b) radiant natural convector

heat exchanger large surface area

convected air

products of combustion

radiants

cool air

(a)

(b)

FIG 11.21 Wall furnace forced convection heater.

The air which is drawn into the unit by the fan does not mix with the products of combustion. This type of unit gives a rapid build-up of air temperature in the room and an even distribution of heat throughout the room. These units are available as a console or inbuilt unit or as a wall furnace. Some units are available with a balanced flue system.

BALANCED FLUE SYSTEMS

See Chapter 9 'Fluing and ventilation'.

COMMISSIONING ROOM HEATERS

Light up and check

1 Check room ventilation.
2 Purge supply line and pilot supply.
3 Check for leaks.
4 Attach manometer, check and set working pressure of appliance with main burner full on.
5 Check and adjust aeration of main burner.
6 With main burner off, check and adjust pilot.
7 When applicable, check operation of the thermostat.
8 Check the appliance consumption against the data plate.

9 Check and time operation of flame failure device.

Installation check

1 Check that the flue is fitted according to the AS 5601 code and local authority codes.
2 Check operation of the flue.
3 Inbuilt room heaters must be secured and sealed to the room as per manufacturer's instructions.
4 Check operation of the fan.
5 Check that the approved double pole isolation switch is fitted or check that there is an accessible three pin plug and switch to isolate electrical power to the unit.

Instructions for customer

1 Instruct customer on correct lighting procedure.
2 Ensure customers understand and have them demonstrate to you the lighting procedure.

MAINTENANCE OF APPLIANCES

To ensure that appliances perform to their maximum capability and efficiency they need to be maintained at regular intervals. By analysing the flue gases you can gain an indication as to the health of the appliance. If it is producing carbon monoxide then you know that there is a serious problem with combustion which must be rectified immediately or the appliance shut down.

Maintenance includes items such as cleaning the combustion chamber, grills and filters, and checking the appliance operation by re-commissioning it as previously discussed. A maintenance schedule should be filled out to record the results of the maintenance with a copy given to the client and one kept by the gasfitter. This is your record of the health of the appliance.

Individual appliances will have specific requirements, and you need to refer to the manufacturer's instruction as well as observing the general guidelines above.

CHAPTER TWELVE

Meters

FUNCTION OF A METER

The function of a gas meter is to measure the volume of gas used. Its accuracy is tested by a government examiner and sealed to indicate it is within the range of accuracy laid down by the relevant state or territory legislation. Location of the meter and the type of meter connections to be used are normally determined by the local gas authority.

General meter information

The badge plate of the meter gives the following information:

1 capacity in cubic metres (m³)

2 cyclic capacity or capacity per revolution—this is the capacity per operation of the dry meters bellows

3 name of manufacturer

4 meter number

5 year of manufacture

6 although not appearing on the badge plate, the repair date when applicable will be stamped in a prominent position.

Testing and degree of accuracy

All meters must be tested by the government gas examiner every ten years. The following is an example of relevant legislative requirements regarding accuracy and allowable pressure variations. As the relevant Acts are state and territory laws; there may be some variations from state to state.

1 The degree of accuracy shall be between 3% slow and 2% fast.

2 The pressure drop across the meter shall not exceed:

0.125 kPa for 1.38 kPa service

0.250 kPa for 2.75 kPa service.

3 Oscillation in the meter's outlet pressure caused by the meter shall not exceed 25 Pa.

TYPES OF METERS

There are two common types of meters: the positive displacement type and the rotary or flow type.

Positive displacement meters

Wet positive displacement meters

This type of displacement meter consists of a drum which rotates on a central shaft. Gas enters the inner drum above the water line and fills one of the four measuring compartments. As gas enters the measuring chamber, it causes the inner drum to rotate and when the chamber is completely filled, no more gas can enter. The gas then passes out of the drum and through to the outlet connection. When gas is passing out of a chamber to be used downstream, this lowers the pressure in that chamber, which allows the higher pressure in the next chamber to rotate the drum, thus ensuring a continuous flow of gas when required at the outlet. The meter must be fitted level, with the liquid level correct. This ensures the high degree of accuracy that the wet meter is renowned for.

The wet meter is no longer used for the recording of gas used by customers due to the problems of setting and maintaining the water level and adding moisture to the gas, but because of its accuracy, it is used for testing purposes in laboratories (Fig 12.1).

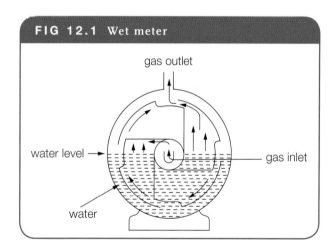

FIG 12.1 Wet meter

gas outlet

water level

gas inlet

water

Dry positive displacement meters

The dry meter consists of three compartments: the upper compartment containing the registering mechanism and the two lower compartments containing a measuring chamber and a bellows in each. The two bellows operate alternately as one inhales gas and the other exhales gas. The two measuring chambers also alternately receive and expel gas, while slide valves control the flow of expelled gas to the meter outlet. A flag arm relays each movement of the bellows to the registering mechanism which records each displacement of gas.

The dry meter is used for all domestic installations and many of the commercial and industrial installations requiring a capacity of up to 75 m³/h (Fig 12.2).

Rotary or flow meter

The rotary meter is ideal for measuring large gas rates and is available with capacities up to 6000 m³/ h. This type of meter is very accurate when measuring large volumes of gas, but it should not be used where the gas rate required may drop below 10% of the meter's maximum capacity.

FIG 12.2 (a) Dry (diaphragm) meter, cutaway view 1

FIG 12.2 (b) Dry (diaphragm) meter, cutaway view 2

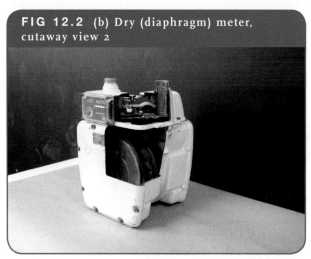

The meter consists of two impellers which rotate inside an oval-shaped cylinder. The impellers are rotated by the pressure of the incoming gas. Each impeller rotates on a shaft which is parallel with the axis of the cylinder. The impellers rotate in such a way that there is always contact with one another, preventing the flow of gas through the centre of the cylinder. A pocket of gas is trapped between the impeller and the cylinder wall but is allowed out at the bottom with each revolution of the impeller. As one impeller expels its pocket of gas, the other impeller fills its pocket of gas and vice versa (Fig 12.3(a) and (b)).

INSTALLATION OF METERS

The location of the gas meter is determined by the local authority, but generally it needs to be located where it is easily accessible and not subject to damage or exposed to detrimental conditions. On domestic premises the meter is usually located at the front of the property either at the boundary or on the corner of the house. The plumber and gasfitter usually runs the consumer piping from the outlet connection of the meter with the utility and consumer service installed by the authority. It is important that the AS 5601 code and any local authority codes are followed when carrying out this work.

The local gas authority determines the type of materials to be used for the meter connections. Domestic meter connections are usually made of hard drawn copper or a similar rigid material that has the capability to support the meter. The riser for connecting to the inlet and outlet of the meter must be vertical, the correct distance apart so as not to put any strain on the meter, and must terminate at the level required by the local authority. Figure 12.4 is an example of a local authority's requirements for fixing a meter. Requirements may vary from one gas utility to another.

FIG 12.3 (a) Rotary meter, (b) detail of inside of rotary meter

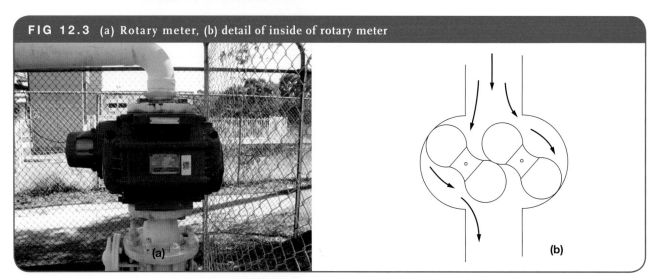

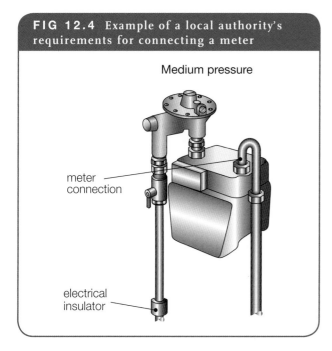

FIG 12.4 Example of a local authority's requirements for connecting a meter

Medium pressure

meter connection

electrical insulator

Types of installations

Master meter

This meter is connected to the inlet service and supplies gas to subsidiary meters.

Subsidiary meter

This meter is not connected directly to the service but it meters gas which has previously passed through a master meter. The local authority may only read and account for the gas passing through the master meter, or it may accept the responsibility for reading and accounting for the gas that is passed through both the master meter and the subsidiary meter.

Note: When installing a meter or adding appliances to an existing installation, the plumber and gasfitter must check that the volume of gas required is within the rated capacity of the meter. Elevated metering pressures effectively increase the capacity of the meter; check with your local authority for metering pressures and capacities.

Connecting a meter to a high or medium pressure inlet service riser

Domestic meters must be connected to a low pressure supply, therefore a medium to low or high to low pressure regulator must be fitted between the meter and the medium or high pressure valve on the service riser.

Bonding straps

If an electrical leakage takes place because of faulty electrical equipment, there is a danger that the gas installation may act as an earth. If the meter was disconnected and separated from the installation, arcing may occur causing any gas present to ignite. To prevent electrical arcing when disconnecting meters or cutting metallic pipes, bonding straps should be connected to the inlet riser and the outlet service or either side of the cut when cutting metallic pipes. The bonding straps should have insulated handles on the clips which are to be attached to the pipes.

When attaching the bonding strap, it must be first connected to the base of the inlet riser and then to the outlet service. Bonding straps must only be used while

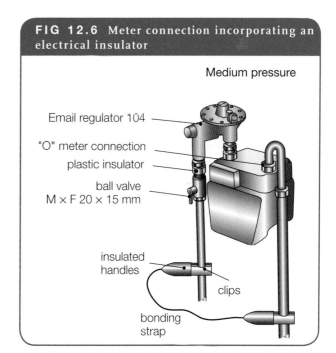

FIG 12.6 Meter connection incorporating an electrical insulator

Medium pressure

Email regulator 104

"O" meter connection

plastic insulator

ball valve
M × F 20 × 15 mm

insulated handles

clips

bonding strap

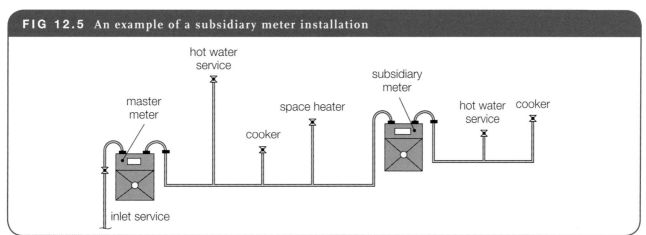

FIG 12.5 An example of a subsidiary meter installation

hot water service

master meter

subsidiary meter

space heater

cooker

hot water service   cooker

cooker

inlet service

disconnecting or changing the meter; once the job is complete, remove the bonding straps.

On metallic mains a plastic bush acts as an electrical insulator which prevents the transfer of electrical potential and arcing. When installing pipes on the outlet of the meter, care must be taken not to make metal to metal contact below the insulator between the inlet riser and the outlet installation (Fig 12.6). Gas pipes must not be used as an earthing point.

Plastic services may be earthed by draping a damp rag over the pipe and onto the ground surface. This will help to discharge any static electricity generated during cutting. Earthing of metallic pipes connected to plastic mains needs special consideration and you may need to consult an electrician for means of earthing.

READING GAS METERS

Method of reading consumer billing meters

Metric meters show the index by using digits. Read from the left-hand side and record the first four digits only. These indicate the number of cubic metres: the fifth digit is of a different colour and indicates tenths of a cubic metre (Fig 12.7). The index shown here should be recorded as 1481, but record whole numbers only. Some meters have subsequent digits that also measure hundredths and thousandths of a cubic metre and usually do not have a test dial as the meter reads these smaller flows (Fig 12.8).

The first four numbers are used to record gas consumed. The last three digits can be used when checking gas rates.

CHECKING GAS RATES

The test dial of a meter is used to check gas rates of appliances or to check the rate of gas escapes. Meters that do not have a test dial have additional dials that measure small amounts of gas. The gas rate of an appliance is meant to be within 10% of its rating on the data plate and this can only be checked by completing the following calculation.

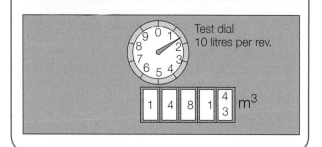

FIG 12.7 Metric meter index reading 1481.3 m³

Test dial
10 litres per rev.

1 4 8 1 $\frac{4}{3}$ m³

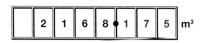

FIG 12.8 Modern metric meter index reading

2 1 6 8 1 7 5 m³

To check the gas rate of an appliance using a test dial of a metric meter which records litres, turn on the appliance, and use a stop watch to record a reasonable number of litres on the test dial (e.g. 20 litres in 37 seconds). The following formula will change the test reading into gas rate per hour.

$$GR = \frac{V \times H \times HV}{T}$$

Where:

- GR = gas rate measured in cubic metres
- V = volume of test measured in cubic metres
- H = number of seconds in one hour
- HV = heating value of the gas expressed in MJ/m³
- T = number of seconds taken to conduct test

Note: Litres will need to be converted to cubic metres by dividing by 1000.

Therefore as there is 1000 litres of gas in one cubic metre then 20 litres would equal 0.02 cubic metres and the gas rate the appliance would be using can be calculated as follows:

$$GR = \frac{V \times A \times HV}{T}$$
$$= \frac{02 \times 3600 \times 38}{37}$$
$$= 73.95 \text{ MJ/hr}$$

Refer to Figure 12.9.

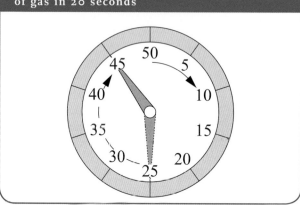

FIG 12.9 Metric test dial measured 20 litres of gas in 20 seconds

To check the gas rate of an appliance using a test dial (0.05 m³) of a metric meter which records cubic metres, turn on the appliance, and use the stop watch to record a reasonable number of cubic metres on the test dial (e.g. 0.05 m³ in 1 minute and 40 seconds). The following formula will change the reading into gas rate per hour:

$$GR = \frac{V \times H \times HV}{T}$$

Where:

- GR = gas rate measured in cubic metres
- V = volume of test measured in cubic metres
- H = number of seconds in one hour
- HV = heating value of the gas expressed in MJ/m³
- T = number of seconds taken to conduct test

Note: Test time will need to be expressed in seconds by multiplying the whole minutes by 60 and adding to the seconds.

Therefore, as 1 minute and 40 seconds equals 100 seconds (60 + 40), the rate at which the appliance is consuming gas can be calculated as follows:

$$GR = \frac{V \times H \times HV}{T}$$

$$= \frac{0.05 \times 3600 \times 38}{100}$$

$$= 68.4 \text{ MJ/hr}$$

To check the gas rate of an appliance using a metric meter without a test dial but which records in cubic metres to three decimal places, turn on the appliance, and use the stop watch to record a reasonable number of cubic metres on the dial (e.g. 0.01 m³ in twenty seconds). The following formula will change the reading into gas rate per hour:

$$GR = \frac{V \times H \times HV}{T}$$

Where:

GR	=	gas rate measured in cubic metres
V	=	volume of test measured in cubic metres
H	=	number of seconds in one hour
HV	=	heating value of the gas expressed in MJ/m³
T	=	number of seconds taken to conduct test

Note: When the last digit rotates a full revolution, the second last digit will have clicked over 1 (e.g. 7 to 8) and measured 0.01 m³.

Therefore the rate at which the appliance is consuming gas can be calculated as follows:

$$GR = \frac{V \times H \times HV}{T}$$

$$= \frac{0.01 \times 3600 \times 38}{20}$$

$$= 57.72 \text{ MJ/hr}$$

Note: The longer the test time the lesser the effect of errors in timing; that is, a one second error will have a greater margin of error on the result of a test conduct over 20 seconds as compared to a test for the same appliance conducted over 100 seconds.

CHAPTER THIRTEEN

Liquefied petroleum gas

Liquefied petroleum gas (LPG) is the name given to a whole group of gases, but those that are marketed throughout Australia for use in homes, farms, industries, commercial premises and portable barbecues are propane and also butane in warmer climates, such as in Queensland. Autogas is a blend of propane and other liquefied petroleum gases like butane, hence the reason why it cannot be used in domestic applications. The most important characteristic of propane is that it can be easily stored as a liquid, making it a highly concentrated source of energy, but it will vaporise rapidly so that it can be burnt as a gas.

CHARACTERISTICS OF PROPANE (C_3H_8)

The boiling point of propane is −42°C. If the plumber and gasfitter is to handle propane effectively and safely, he or she must understand boiling, the effects of temperature on pressure within the cylinder and the vaporisation rate of liquid propane.

Before considering the boiling of propane, let us consider the boiling of a liquid we are all familiar with, that of water. It is generally accepted that if water is heated, it will boil at 100°C. If water is placed in an open container and heat is applied to the container, not only will the liquid expand but the molecules of H_2O within the liquid will expand and cause a pressure build-up within the liquid. The atmospheric pressure on the surface area of the liquid is preventing boiling from taking place, but when the water reaches a temperature of 100°C the pressure within the liquid will be greater than the atmospheric pressure on the surface. This allows the liquid to expand and form gaseous bubbles which force themselves out of the liquid (Fig 13.1(a) and (b)).

It can be seen that boiling takes place when the pressure within the liquid is greater than the pressure exerted on the surface of the liquid. Therefore, boiling can be retarded by increasing the pressure on the surface of the liquid. An obvious example of this is the radiator of

a car. The radiator cap allows the pressure on the surface of the liquid to build up and thus prevent the water from boiling at 100°C, but if the cap is removed when the water is above 100°C, the additional pressure on the surface of the liquid is lost. Because the pressure within the liquid is greater than atmospheric pressure, rapid boiling takes place and forces water and steam out of the radiator.

If liquid propane is placed in a sealed container and the container is subjected to an air temperature of 20°C, which is 62°C above the atmospheric boiling point of propane, heat will be conducted through the metal sides of the container into the liquid propane. The propane will boil rapidly but the propane vapour will be trapped above the surface of the liquid. As the propane continues to boil, the vapour will become compressed, building up pressure on the surface of the liquid. When the vapour pressure on the surface of the liquid is equal to the pressure within the liquid, a state of equilibrium will exist and boiling will be retarded. If the gas is drawn off from the top of the cylinder, the drop in pressure will allow the liquid to boil to replace it (Fig 13.2(a) and (b)).

FIG 13.2 Vaporising of propane: (a) vapour pressure P_1 = pressure within liquid P_2; (b) P_1 is less than P_2

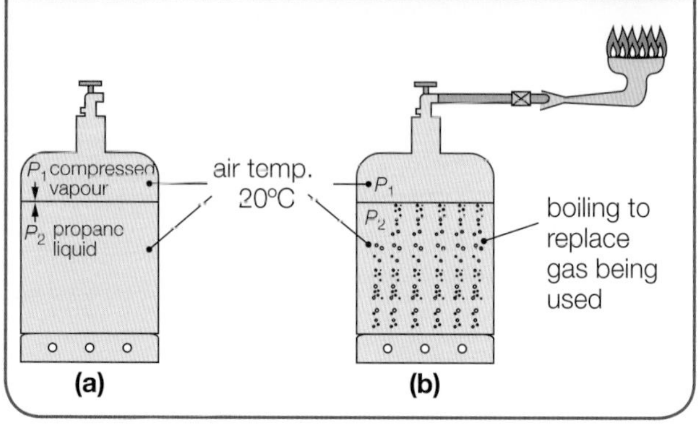

P_1 compressed vapour
P_2 propane liquid
air temp. 20°C
P_1
P_2
boiling to replace gas being used

(a)   (b)

FIG 13.1 Boiling principle

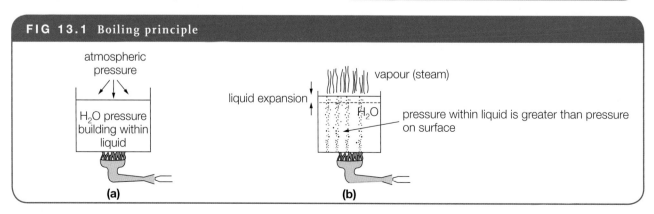

atmospheric pressure

H_2O pressure building within liquid

(a)

liquid expansion

vapour (steam)

H_2O

pressure within liquid is greater than pressure on surface

(b)

EFFECTS OF TEMPERATURE ON PRESSURE

In the same way that the amount of heat applied to a kettle will determine how fast the water in the kettle will boil, so will the temperature surrounding the cylinder determine how much boiling is going to take place inside a propane cylinder. The higher the temperature surrounding the cylinder, the greater the pressure will have to build up on the surface of the liquid to retard boiling. An example of the effects of temperature on pressure can be seen from the following:

Temperature	Vapour pressure
0°C	372 kPa
15°C	627 kPa
40°C	1247 kPa

Propane cylinders are pressure vessels capable of withstanding pressure well above those illustrated above, but cylinders should not be exposed to temperatures other than normal atmospheric air temperatures.

VAPORISATION

Vaporisation rates

The vaporisation rate of propane is determined by the following:

1   volume of liquid in the cylinder (wet area of cylinder);

2   temperature to which the cylinder is exposed.

It is important that the volume of gas being drawn off from the cylinder is not in excess of the vaporisation rate available from the energy being conducted through the cylinder from the surrounding air. As the level of the liquid drops, the area that surrounds it is less (wet area) which decreases the rate of vaporisation due to less heat energy being available. Table 13.1 indicates the vaporisation rate of a 45 kg domestic cylinder under varying conditions.

Icing up of cylinders

The liquid propane in a cylinder absorbs energy from the air surrounding the cylinder. If gas is drawn off faster than the air surrounding the wet area can vaporise it, the energy contained within the liquid will be used to assist with the vaporisation. This will cause the temperature of the liquid propane to drop. When it reaches the temperature of the dewpoint of the air surrounding the cylinder, condensation will form on the cylinder. If the excessive draw-off of gas continues, the temperature of the liquid will continue to drop, causing the condensation on the cylinder to ice up. If icing up of a propane cylinder occurs, consideration

should be given to increasing the size of the cylinder or tank, thus increasing the available wet area or connecting more cylinders and tanks, again increasing the wet area. The alternative to using a larger cylinders or more cylinders is to use a vaporiser.

Vaporisers

Although vaporisers will provide additional energy to meet the vaporisation rate required, consideration must be given to the capacity of the tank and the availability of refills.

Atmospheric vaporisers

Atmospheric vaporisers consist of a finned tube heat absorber which effectively increases the wet area of the tank, thus absorbing sufficient heat from the atmosphere to vaporise sufficient liquid to meet the demands of the appliances.

Direct fired vaporiser

These vaporisers usually use a gas burner to supply heat to the vaporiser vessel. The burner is of the automatic type which operates when the vapour pressure falls due to the draw-off from the storage tank.

Indirect fired vaporiser

The heating medium is in the form of hot water or steam supplied to the heat exchanger. Constant vapour pressure is maintained by automatic control of the heating medium.

Expansion rate of propane

When liquid propane changes from a liquid to a gas, it expands 273 times. Therefore, 1 L of liquid expands to form 273 L of gas.

Liquid expansion rate of propane

If the temperature of liquid propane was raised from 10 to 30° C, the liquid volume would have increased by 6%. If cylinders were completely filled, any increase in temperature would cause a rapid build-up of pressure in the cylinder with the probability of the pressure relief valve venting the excess pressure to atmosphere. Space must be left in the top of the cylinder for liquid expansion. The maximum fill level for propane cylinders is 83%, the remaining 17% being for liquid expansion and vapour pressure.

Note: The additional properties and characteristics of propane are contained in Chapter 1.

TABLE 13.1 Vaporisation rate of 45 kg domestic cylinder			
Temperature	Maximum fill level	50% of maximum fill level	10% of maximum fill level
	45 kg	22.5 kg	4.5 kg
21°C	316 MJ/h	176 MJ/h	69 MJ/h
16°C	285 MJ/h	162 MJ/h	63 MJ/h
4.5°C	225 MJ/h	131 MJ/h	52 MJ/h

DECANTING PROCEDURES

Following the correct procedure when decanting is undoubtedly a most important safety factor when handling propane. Overfilling of cylinders has been a major factor in the majority of serious LPG accidents.

Decanting is the transfer of liquid from cylinder to cylinder or tank to cylinder without the use of a pump. Cylinders of 9 kg or less are generally filled by decanting from a blue top liquid withdrawal cylinder. The correct fill level of 83% is ensured by filling to the level of the fixed dip tube (Fig 13.3(a) and (b)).

Note: When liquid propane is vented to atmosphere through the bleeder valve, the temperature will drop to approximately −42°C, so gloves must be worn at all times to prevent cold burns.

The filling of larger cylinders which do not have a fixed dip tube can only be carried out on an approved filling platform using approved equipment and scales to ensure correct filling level.

CYLINDERS AND PRESSURE RELIEF VALVES

Construction of cylinders

Cylinders are constructed in accordance with the Australian Standards for gas cylinders and tested also to the relevant Australian Standard/s.

Pressure relief valves

To ensure pressures are contained to safe limits, pressure relief valves are fitted to the vapour section of all LPG cylinders and tanks.

Small portable cylinders use a fusible plug fitted into the bleeder valve. The fusible plug will blow out of its housing should the temperature reach 100°C or the pressure reach 2585 kPa, and all the contents of the cylinder will probably be lost.

Larger portable cylinders use a spring-loaded pressure relief valve. If the pressure inside the cylinder reaches 2585 kPa, the relief valve will open allowing the excess pressure to vent. When the pressure drops to a predetermined level, the pressure relief valve will reseal. The most common type of pressure relief valve used in portable cylinders is incorporated in the control valve (Fig 13.5).

Most tanks use a separate pressure relief valve fitted direct to the vapour section of the tank. The pressure relief valve on a tank is set to relieve the pressure should it reach 1725 kPa. An internal or external type of pressure relief valve is available for tanks (Fig 13.6(a) and (b)).

On large tanks, gas discharging from pressure relief valves may be carried to a safe distance above the tank through metal pipe extensions.

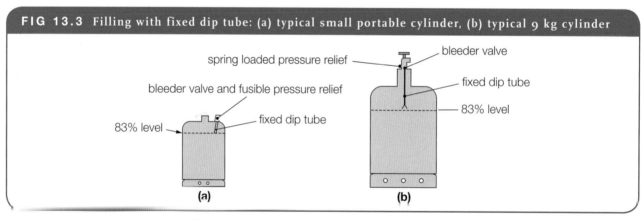

FIG 13.3 Filling with fixed dip tube: (a) typical small portable cylinder, (b) typical 9 kg cylinder

spring loaded pressure relief

bleeder valve and fusible pressure relief

83% level

fixed dip tube

(a)

bleeder valve

fixed dip tube

83% level

(b)

FIG 13.4 LPG cylinder, cutaway view to show fill level gauge

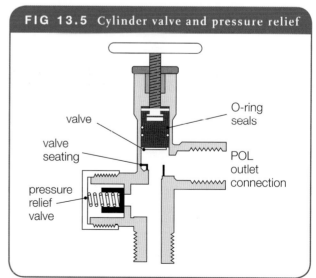

FIG 13.5 Cylinder valve and pressure relief

valve

valve seating

pressure relief valve

O-ring seals

POL outlet connection

FIG 13.6 (a) Internal type pressure relief valve, (b) external type pressure relief valve

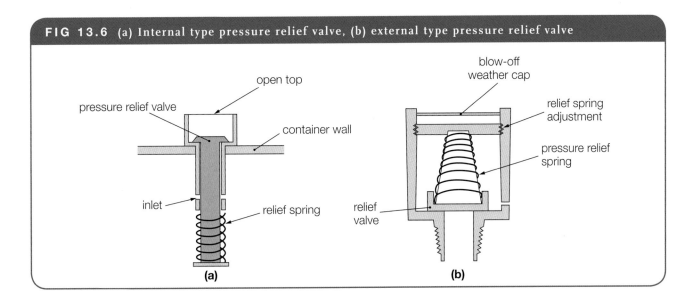

(a)   (b)

TRANSPORTATION OF PORTABLE CYLINDERS

To ensure the safe transportation of portable cylinders, the following points should be observed:

1 Transport cylinders in an upright position to ensure the pressure relief valve will vent vapour and not liquid if an over-pressure situation should arise.

2 Use an open utility or truck.

3 Secure cylinder to the vehicle.

4 Do not carry by cylinder valve.

5 A suitable protective cap must be fitted over the cylinder valve unless casing securely attached to the cylinder body offers suitable protection to the valve.

6 When loading heavy cylinders on your own, unload onto soft materials (e.g. old car tyres).

INSTALLATION METHODS

Domestic twin-cylinder installations

The twin-cylinder installation consists of the following components.

Regulator

There are two types of regulators commonly in use on twin-cylinder installations: a manual changeover type and an automatic changeover type. The automatic changeover regulator has the advantage of supplying an uninterrupted supply of gas. As soon as the pressure in the cylinder that the gas is being drawn from drops below 7 kPa, the regulator will automatically draw gas from the other cylinder. A red disc indicates when the empty cylinder needs changing.

Operation of the automatic changeover regulator

1 Set indicator on top of regulator to point in the direction of the cylinder from which the gas is to be drawn.

2 The cam will increase the loading on the first stage of the regulator to 80 kPa.

3 The pressure of gas from the cylinder will be reduced to 80 kPa as it passes through the first-stage regulator.

4 The 80 kPa pressure will be allowed to pass to the underside of the diaphragm of the first-stage regulator which controls the flow of gas from the second cylinder. The loading on the top side of the diaphragm is 40 kPa; therefore, the diaphragm will lift and shut off the supply of gas from the second cylinder.

5 When the pressure of gas in the cylinder from which gas is being drawn off drops below 40 kPa, the loading on the top side of the diaphragm on the first-stage regulator of the second cylinder will open the valve and allow gas to be drawn off.

6 The red disc will indicate to the customer that the first cylinder is empty.

7 Gas passes from the first-stage regulator at either 80 kPa or 40 kPa to the second-stage regulator where it is reduced to the normal working pressure of 2.75 kPa (Figures 13.7(a) and (b)).

Operation of regulator with manual changeover valve

1 The valve on top of the manual changeover valve is turned to point in the direction of the cylinder from which gas is to be drawn.

2 The cam pushes the valve off the seating of the cylinder that is to be used and closes off the back-up cylinder.

3 When the cylinder is empty of gas, the customer must manually turn the valve to draw gas from the reserve cylinder (Figures 13.8(a) and (b)).

Pigtails

This is a flexible copper connection which joins the cylinder to the pressure regulator. An expansion loop is fitted to permit the flexibility required by expansion and contraction or slight movement of the cylinder as well as facilitating manual changeover of cylinders.

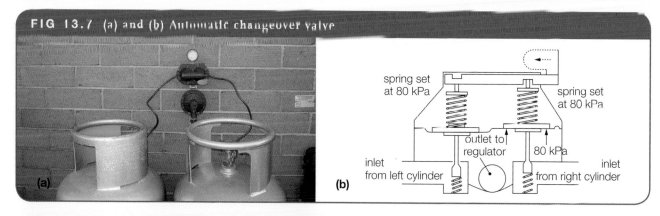

FIG 13.7 (a) and (b) Automatic changeover valve

(a)

(b)

spring set at 80 kPa

spring set at 80 kPa

outlet to regulator

80 kPa

inlet from left cylinder

inlet from right cylinder

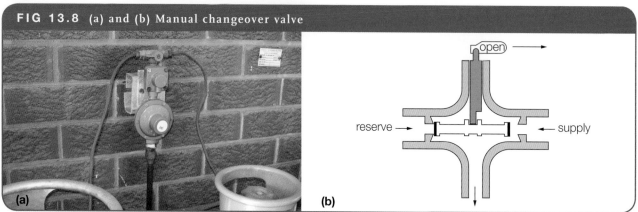

FIG 13.8 (a) and (b) Manual changeover valve

(a)

(b)

open

reserve

supply

Combined cylinder and pressure relief valve

The cylinder valve which permits the cylinder to be turned on or off also consists of a spring-loaded pressure relief valve which serves as a safety relief should the cylinder pressure exceed 2585 kPa. The pressure relief valve must be pointing away from the building.

Canopy

The canopy is fitted to protect cylinder valves and regulators from the weather and damage. When fitting the canopy, ensure it is above the level of the pressure relief valve. Under no circumstances must it impede or redirect any flow from the pressure relief valve.

Base

The cylinders are mounted on a non-corrosive or fire-resistant base, usually a concrete slab fitted 50 mm above the ground level. This base also prevents the cylinders from subsiding, which could cause the soft copper pigtail connections to strain or fracture.

Installation pipe to appliances

The installation pipe to the appliance should be sized and fitted in accordance with the AS 5601. See Chapter 10 for method of using pipe-sizing charts. Refer to Figure 13.9.

Location of cylinders

Cylinders and first-stage regulation equipment must be located outside buildings and comply with AS 5601. Particular attention is drawn to the distance from ignition sources and openings into buildings—refer to AS 5601.

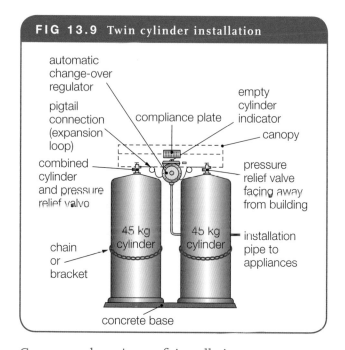

FIG 13.9 Twin cylinder installation

automatic change-over regulator

pigtail connection (expansion loop)

compliance plate

combined cylinder and pressure relief valve

empty cylinder indicator

canopy

pressure relief valve facing away from building

45 kg cylinder

45 kg cylinder

chain or bracket

installation pipe to appliances

concrete base

Caravan and marine craft installations

Caravan and marine craft installations must be made in accordance with the AS 5601 and any state laws that apply to caravan and marine craft installations. Cylinder locations for a marine craft are shown in Figure 13.11.

Cylinder or compartment drain shall not be less than 1 m from any opening that is below it, not less than 150 mm from any opening that is above it, and not less than 2 m from any fixed source of ignition. All distances should be exceeded where practicable.

FIG 13.10 (a) and (b) Exchange cylinder location (see AS 5601)

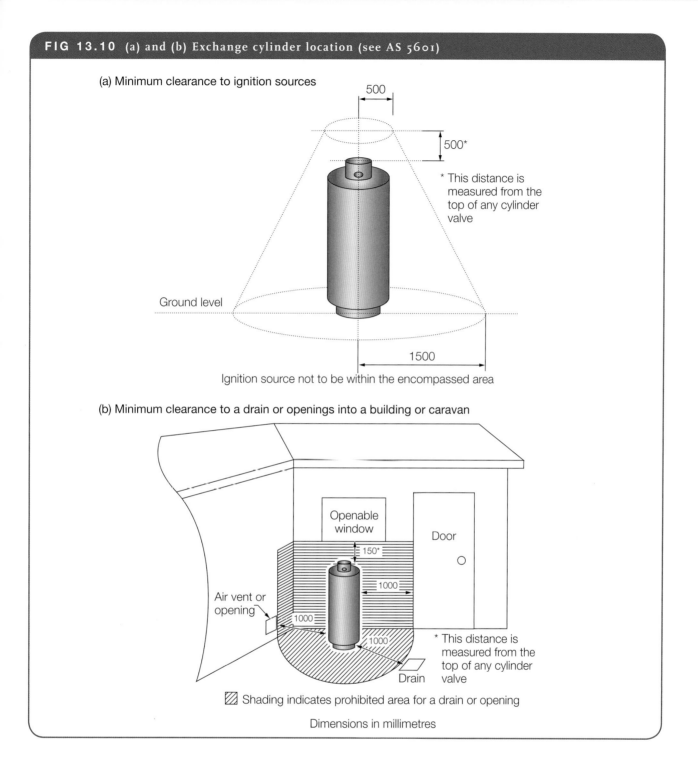

(a) Minimum clearance to ignition sources

500

500*

* This distance is measured from the top of any cylinder valve

Ground level

1500

Ignition source not to be within the encompassed area

(b) Minimum clearance to a drain or openings into a building or caravan

Openable window

Door

150*

1000

Air vent or opening

1000

1000

* This distance is measured from the top of any cylinder valve

Drain

⬚ Shading indicates prohibited area for a drain or opening

Dimensions in millimetres

Multiple cylinder installations

If a twin cylinder installation cannot supply sufficient gas, a multiple cylinder installation may be used. Two banks of cylinders usually consisting of two cylinders per bank are normal. Anything in excess of this should be with the permission of the local technical regulator or authority.

An automatic or manual changeover regulator can be used, but check to ensure it can cope with the demands of the system. The manifold can be made from ready-made tee blocks which allow for a pigtail connection into the centre and a POL connection for extending to the next cylinder (Fig 13.12).

An alternative is to manufacture a manifold from approved quality copper tube and fittings, or welded steel and fittings.

Tank installations

The installation of tanks for the supply of LPG shall comply with AS 1596. The statutory authority for tank installations is the Department of Labour, which must be informed before installation.

FIG 13.11 Location of cylinders on marine craft

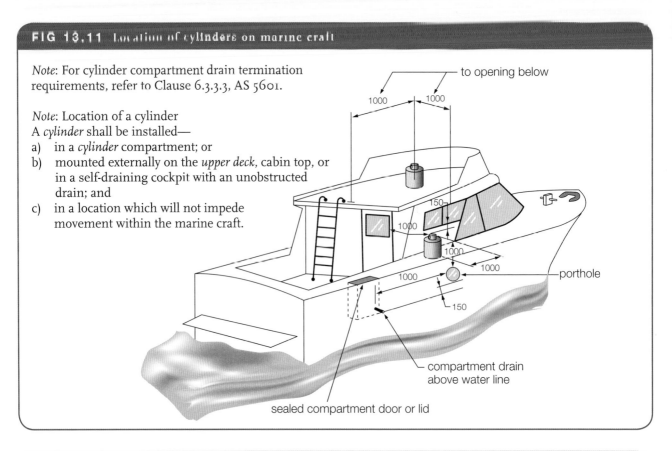

Note: For cylinder compartment drain termination requirements, refer to Clause 6.3.3.3, AS 5601.

Note: Location of a cylinder
A *cylinder* shall be installed—
a) in a *cylinder* compartment; or
b) mounted externally on the *upper deck*, cabin top, or in a self-draining cockpit with an unobstructed drain; and
c) in a location which will not impede movement within the marine craft.

FIG 13.12 Multiple cylinder installation using tee blocks

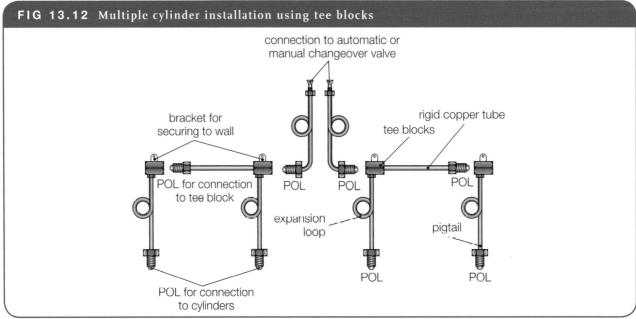

Equipment commonly used on a tank installation
Refer to Figure 13.13.

1 Liquid filling valve. This is the valve to which the tanker's liquid hose is attached to fill the tank with liquid propane.

2 Service line valve. This is the valve through which gas is drawn from the storage tank to the pipe installation.

3 Stage one regulator. This regulator is fitted on the outlet side of the service line valve and reduces the tank pressure to a suitable operating pressure.

Note: Service line operating pressures are usually set to operate at 70–84 kPa. The operating pressure is determined in design after considering load factors, length and diameter of run, future gas requirements and the range of ambient air temperatures to which the service line is subjected.

4 Expansion loop. This is fitted to permit movement of pipework due to expansion and contraction or movement of the tank settling down. In this way, fractures or damage to the installation and equipment are avoided.

FIG 13.13 LP tank installation

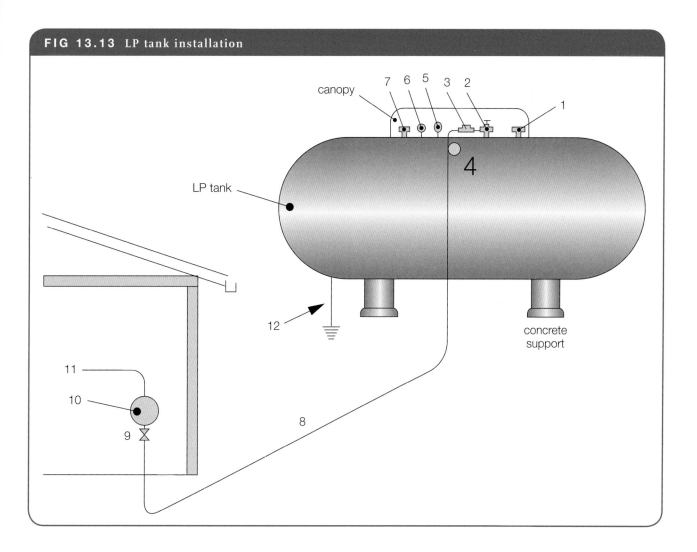

5 Liquid level gauge. This indicates the liquid content of the tank.

6 Pressure gauge. This indicates the vapour pressure in the tank.

7 Pressure relief valve. The safety requirement is fitted in the vapour section and set to discharge at 1725 kPa.

8 Medium pressure line. It is standard practice to run a high pressure line from the tank position to the building. This permits small diameter pipe to be used between the first- and second-stage regulator.

9 Valve. This valve permits the isolation of the medium pressure line from the second-stage regulator.

10 Stage two regulator. The stage two regulator is fitted at the building and reduces the medium pressure to a low pressure to operate the appliances.

Note: Stage two regulators incorporate a pressure relief section and therefore must be fitted outside the building.

11 Internal installation. Installation pressure is 3 kPa. The internal pipes are sized in accordance with the AS 5601 which allows a maximum of 250 Pa pressure drop from the second-stage regulator

to any appliance to ensure an appliance working pressure of 2.75 kPa.

12 Tank earthing. Tanks over a specified size have to be earthed to discharge any static electricity that may build up and possibly cause ignition.

Sizing of the medium pressure line on an LPG tank installation

Sizing the medium pressure line between the first- and second-stage regulator can be done by using the table in AS 5601. First determine the length of line (index length), operating pressure and volume of energy in MJ/h required. Refer to Chapter 10 for further information on pipe sizing. Note that, before installing any tank, you must check the AS 1596 code. Compliance with this code is mandatory. Inform the statutory authority prior to location and installation of tank.

BURIAL METHOD OF LPG TANK INSTALLATION

The burial method of LPG tank installation is widely used in both industrial and domestic situations. The benefits include a more efficient and less obtrusive use of space, a safer and more secure location; and regular monitoring of volume levels, with appropriately scheduled refills, without the necessity of changing over cylinders. Fig 13.14 shows a large tank being installed in a petrol station. In both industrial and domestic settings the tank is typically coated, with cathodic protection (most commonly using magnesium or zinc anodes) providing backup for any coating deterioration. Cathodic protection is an electrical method of preventing corrosion of metallic structures (most commonly iron and steel, including stainless steel) situated in electrolytes (usually soil or water). The surface of the tank should never be in contact with gravel or rocks or any rough backfill which could damage the protective coating. Sand or firm earth free of rocks and abrasives should be used to cover a tank after installation. A canister

may be used for extra cylinder protection. Fig 13.15 shows a protective polyethylene canister housing a gas cylinder for domestic use. It is essential that an LPG tank never be installed under a vehicular traffic area, for example in a driveway. The tank may need to be anchored in concrete to prevent floating, which may occur if water rises to a higher level than the gas in the tank.

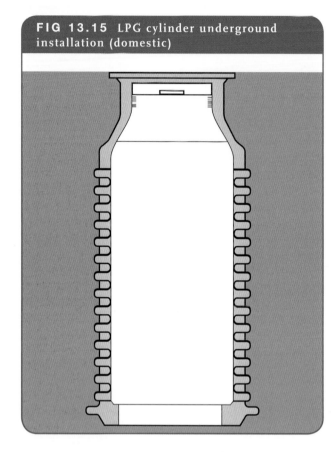

FIG 13.15 LPG cylinder underground installation (domestic)

FIG 13.14 LPG tank underground installation (industrial)

CHAPTER FOURTEEN

Safety in the gas industry

The gas industry's standards and codes are set to prevent hazardous situations from arising, but from time to time things do go wrong. It is important that plumbers and gasfitters are capable of assessing the situation and taking appropriate action.

We do not know how we will act in an emergency situation until the time arises. However, training in handling hazardous situations does increase the likelihood that correct decisions will be made. A step-by-step account of how to handle a particular hazardous situation is not possible because no two accidents are completely alike, but the factors indicated in this chapter have proven value.

HAZARDS

The hazards of escaping unburnt gas are:

1 fire

2 explosion

3 asphyxiation.

FIG 14.1

The hazards of partially burnt gas are:

1 asphyxiation

2 carbon monoxide poisoning.

HANDLING A GAS-AFFECTED BUILDING

If the plumber or gasfitter has reason to believe that the presence of gas in a building may be sufficient to cause any of the hazards listed above, then he or she should consider the following factors in determining their course of action.

1 Assess danger by looking for ignition sources and checking for excessive gas around isolation valves. Consider possible need for breathing apparatus.

2 Turn off the supply of gas to the building. Do not assume that turning off at the meter inlet will stop the escape—the source could be a nearby main or service. A test of the internal installation will determine whether the source is inside or outside the building.

3 If you consider it necessary, evacuate the building. To achieve an orderly evacuation you must remain calm— imagine how you would feel if a firefighter directing the evacuation of a fire-affected building was in a state of panic.

4 Eliminate ignition sources as follows:

   (a) Extinguish any naked flames.

   (b) Prevent smoking and striking of matches.

   (c) Do not carry matches or lighters with you into the affected area.

   (d) Do not touch any electrical switches; if lights are on, leave them on. Vehicle ignition systems are also a source of ignition.

   (e) Thermostatically or time-controlled equipment may switch on automatically. Isolate the electrical supply in a remote safe area such as the main switch at the meter. If the electrical supply cannot be isolated at the main switch, contact the electrical supply company.

   (f) Use a leak-proof torch. Switch it on in a safe atmosphere and leave it on.

   (g) Although combustible gas detectors are leak proof, wherever possible switch them on in a gas-free atmosphere.

5 Ventilate the building by opening all doors and windows. If the concentration of gas is above the upper explosive limit, somewhere on the perimeter of the affected area an

explosive mixture will exist. Ventilation will help to dilute this mixture.

6   Notify the gas supplier. If you consider some of the other factors to be of higher priority, direct someone else to do this for you. The gas supplier may be the only person capable of turning off the source of gas.

7   Do not enter the affected area unless wearing appropriate breathing equipment, harness and safety line. Extra caution is required if entering in darkness or going into cellars. Do not take risks by entering places you may not easily be able to get out of.

8   Check adjacent buildings to ensure escaped gas is not present in them.

HANDLING LPG EMERGENCIES

Escapes at cylinders on fire

If possible, determine the length of time the cylinder has been on fire and whether flames are in contact with the vapour section of the cylinder. The intense heat may cause the metal to soften. The build-up of pressure in the cylinder can cause a hot spot to bulge and split, causing a BLEVE (boiling liquid expanding vaporising explosion).

If a cylinder were to explode in this way (BLEVE), people in the area would be at risk from fire and flying metal. Therefore, the utmost care must be used when dealing with cylinders that are exposed to fire. Follow this procedure when handling a fire that is in contact with a cylinder:

1   Evacuate the area.

2   Notify the fire brigade. They will be able to approach the cylinder behind a spray from a fog nozzle, and then spray water onto the cylinder with a hose from a safe distance.

If the pressure relief valve is blowing and the gas is on fire but not in contact with the cylinder, the following course of action is recommended:

1   Check to ensure there is no unburnt gas in the area. Spray water onto the top section of the cylinder, taking care not to extinguish the fire from the vented gas. Continue spraying until the cylinder cools and the pressure relief valve reseals.

2   When the pressure relief valve reseals, turn off the gas, disconnect and remove the cylinder. Notify the gas supplier.

Plumbers and gasfitters should not attempt to handle LPG tank fires as this requires equipment capable of supplying copious volumes of water and specially trained personnel. In the event of a tanker fire, evacuate at least 300 m radius from the tank.

Notify the fire brigade and other appropriate authorities.

The Australian LPG codes of practice have ensured the safe use of this fuel for over thirty years. Plumbers and gasfitters can help maintain this excellent safety record by following the codes and Australian Standards when installing cylinders and appliances.

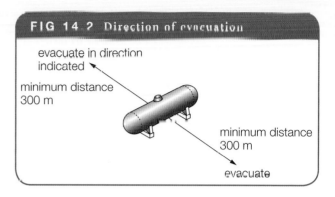

FIG 14.2   Direction of evacuation

evacuate in direction indicated

minimum distance 300 m

minimum distance 300 m

evacuate

CARBON MONOXIDE POISONING AND ASPHYXIATION

Effects of carbon monoxide

When inhaled, carbon monoxide combines with haemoglobin and is pumped around the body. This prevents the blood from carrying oxygen and therefore prevents oxygen from reaching the body tissues. Small concentrations of carbon monoxide can be lethal; for example, one part carbon monoxide in one thousand could be fatal. When towns gas was in use it contained a very high percentage of carbon monoxide and was therefore potentially lethal in its unburnt state. Incomplete combustion with any gas-burning appliance may also produce lethal doses of carbon monoxide, hence the importance of locating, installing and commissioning appliances correctly.

Symptoms of carbon monoxide poisoning

Some or all of the following symptoms may be present in a person suffering from the effects of inhaling carbon monoxide:

1   giddiness

2   lack of control of the muscles

3   shortness of breath

4   semi-consciousness

5   a sense of wellbeing and an insistence that they are all right, even though the above symptoms are evident

6   lips, nose, ears and cheeks becoming a bright cherry red colour.

Effects and symptoms of asphyxiation

If non-toxic gases escape and build up in an area, they do so at the expense of air. The exclusion or reduction of oxygen in an area can have a very quick effect on any person in that area. Asphyxiation is the lack of oxygen in the blood. Too many people in the industry have become complacent about working in areas affected by gases which are not toxic. Asphyxiation has caused deaths in the industry, all of which could have been avoided by the wearing of suitable breathing equipment or waiting until the areas were ventilated.

The symptoms of asphyxiation are:

1   faintness

2   weakness

3 partial or complete lack of consciousness

4 a sense of wellbeing

5 lack of ability to do simple tasks competently and an aggressive reaction when this is pointed out

6 lips and cheeks becoming blue in colour

7 all the facial features becoming a blue-grey colour—the person may be unconscious at this stage.

RESUSCITATION

Cardiopulmonary resuscitation (CPR) is a first-aid technique used to keep victims of cardiac (heart) or pulmonary (breathing) arrest alive until emergency response professionals arrive. The purpose of CPR is to keep oxygenated blood flowing through the body to delay tissue death and extend the opportunity for successful resuscitation without permanent brain damage. A defibrillator is usually needed to restart the heart to normal function.

FIG 14.3 Cardiopulmonary resuscitation (CPR)

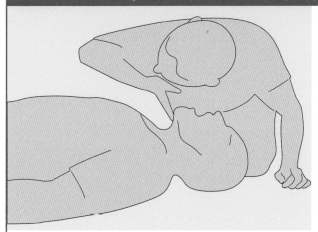

1 Place the victim on his or her back, and tilt the victim's head well back to ensure a clear airway to the lungs and check for breathing and response.

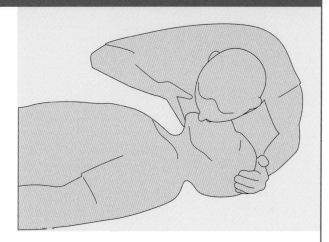

2 Give two (2) breaths into the victim's lungs, trying to obtain a good seal between your mouth and the victim's face.

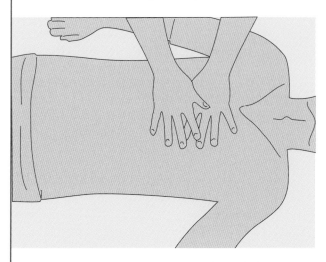

3 Position the hand on the centre of the victim's chest.

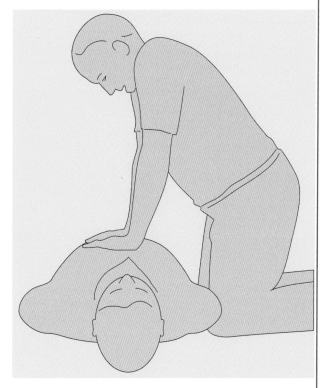

4 Firmly push down 30 times to compress the chest then followed by two (2) breaths.
   Continue to repeat the cycle until professional help arrives. The rate of compressing the chest is 100 per minute or almost 2 per second

Glossary

air, ambient generally, the air surrounding an object

air, recirculated return air passed through the conditioner before being resupplied to the conditioned space

air, reheating of in an air-conditioning system, the final step in treatment if the temperature is too low

air, return see air, recirculated

air, secondary air for combustion supplied to the burner to supplement the primary air

air changes a method of expressing the amount of air leakage into or out of a building or room in terms of the number of building volumes or room volumes exchanged

air conditioning, comfort the process of treating air to control simultaneously its temperature, humidity, cleanliness and distribution to meet the comfort requirements of the occupants of the conditioned space

anemometer an instrument for measuring the velocity of air

appliance a device which uses gas to produce flame, heat, light, power or special atmosphere

appliance flueway a port or passage conveying flue gases within the appliance

appliance flueway terminal a protective device fitted to the exit of an appliance flueway of a flueless appliance

appliance regulator a device fitted to an appliance to control the gas pressure or gas volume delivered to that appliance

approved acceptable to, and meeting the prescribed standards of, the authority having jurisdiction

atmospheric burner a burner where all the air for combustion is provided at atmospheric pressure and is introduced by the inspirating effect of an injector and/or by the natural draught in the combustion chamber

authorised agent (authorised dealer) a person, persons or company appointed to act for and on behalf of a gas merchant

authorised installer a person appointed, licensed, authorised or otherwise permitted by the authority to carry out gasfitting work

authority the authority having jurisdiction or such authority as delegated

automatic burner a burner system which, on starting, follows a self-acting sequence that has been manually or automatically initiated, to provide gas and ignition to the burner without any intermediate manual operation

automatic ignition the lighting of gas at a burner without a manual operation whenever gas flows from the burner

automatic operation the use of a sequence of operations which, once initiated, does not require any intermediate manual operation

bain marie an appliance which keeps cooked food hot, using water or air, in buckets recessed in a flat surface

balanced flue appliance an appliance which has combustion air ducted from, and combustion air ducted to, a common terminal assembly

bayonet fitting a bayonet-style plug and mating socket designed for attachment so that the gas is turned on only when the plug is inserted and secured in a lock-fit recess

bimetallic element an element formed of two metals having different coefficients of thermal expansion, used as a temperature control device

blower a fan used to force air under pressure

boiler a closed vessel in which a liquid is heated or vaporised

boiler, firetube a boiler with straight tubes, which are surrounded by water and steam and through which the combustion products pass

boiler, watertube a boiler in which the tubes contain water and steam, the heat being applied to the outside surface

boiling point the temperature at which the vapour pressure of a liquid equals the absolute external pressure at the liquid-vapour interface

boiling table a commercial appliance intended for the heating of cooking vessels from beneath

burner, atmospheric a gas burner in which air for combustion is entirely supplied by natural draught and the inspirating force created by gas velocity through orifices

burner port the opening in a burner through which gas or an air-gas mixture issues to be ignited and burned

burner unit a number of burners, fed from the same manifold, controlled together, having one supervised ignition source and performing as a single burner

bypass screw an adjusting screw for setting the flow through a bypass

calibration the determination of the relationship between the measured or indicated value of a parameter and its true value

canopy a fixture or hood which serves to remove cooking odours and may be part of the flue gas collection and disposal system, placed above the appliance/s and discharging into the open air by means of convection currents

casing (sleeve) a pipe or duct through which is installed a smaller pipe carrying gas

Celsius (Centigrade) a thermometric scale in which the freezing point of water is 0° and its boiling point 100° at normal atmospheric pressure

change of state change from one phase, such as solid, liquid or gas, to another

chimney effect the tendency of air or gas in a duct or other vertical passage to rise when heated due to its lower density in comparison with that of the surrounding air or gas; in buildings, the tendency toward displacement (caused by the difference in temperature) of internal heated air by unheated outside air due to the difference in density of outside and inside air

circulator a water heater in which water passes to a storage tank after heating

closed top boiling table a boiling table which has a smooth metal heating surface

combination gas control an assembly of two or more different control functions in a single body

combination relief valve a valve which functions in response to both excessive temperature and pressure

combustible construction materials made of or surfaced with wood, compressed paper, plant fibres or other materials that will ignite and burn

combustion products constituents resulting from the combustion of a fuel with oxygen, including the inerts associated with the fuel and the oxygen but excluding any other diluent or contaminant

common flue a flue conveying the flue gases from two or more appliances

condensate liquid formed by condensation of a vapour

conductor, thermal a material which readily transmits heat by means of conduction

continuous cleaning surface the finish on an oven interior which, in the course of ordinary use, does not accumulate any stains or deposits

control, flame safeguard a system for sensing the presence or absence of flame indicating, alarming or initiating control action

control, limit an automatic safety control responsive to change in liquid level, pressure or temperature or position for limiting operation of the controlled equipment

convection heat transfer by the movement of fluid

convection, forced convection resulting from forced circulation of a fluid, as by a fan, jet or pump

convection heater a space heater which produces mainly warm air

convection, natural circulation of gas or liquid (usually air or water) due to differences in density resulting from temperature changes

cooling medium any substance whose temperature is such that it is used, with or without a change of state, to lower the temperature of other bodies or substances

corrosive having a rusting or chemically destructive effect on metals (occasionally on other materials)

cross lighting lighting of one burner from another either directly or by means of an intermediate flame

customer piping (fitting line) that portion of piping which conveys gas from the outlet of the master meter or, in the case of LP gas, from the container regulator outlet to the appliances

cut-off the function of isolating the upstream pressure from the downstream side of a valve or regulator

dehumidification condensation of water vapour from air by cooling below the dewpoint

dehumidifier 1. an air cooler or washer used for lowering moisture content of the air passing through it; 2. an absorption or adsorption device for removing moisture from air

dehydration 1. removal of water vapour from air by the use of absorbing or adsorbing materials; 2. removal of water from stored goods

density ratio of the mass of a specimen of a substance to the volume of the specimen; the mass of a unit volume of a substance

design working pressure maximum allowable working pressure for which a specific part of a system is designed

dewpoint temperature the temperature at which condensation of water vapour in a space begins for a given state of humidity and pressure as the vapour temperature is reduced; the temperature corresponding to saturation (100% relative humidity) for a given absolute humidity at constant pressure

differential of a control, difference between cut-in and cut-out temperatures or pressures

diffuser, air a circular, square or rectangular air distribution outlet, generally located in the ceiling and comprised of deflecting members discharging supply air in various directions and planes and arranged to promote mixing of primary air with secondary room air

direct fired oven an oven in which the products of combustion flow through the oven compartment

direct ignition device a device which provides ignition of a burner without the use of another flame

direct service meter a meter connected directly to the service pipe

drain cock a valve in the bottom of the water vessel, through which the water may be drained from an appliance

draught a current of air, when referring to the pressure difference which causes a current of air or gases to flow through a flue, chimney, heater or space; or when referring to a localised effect caused by one or more factors of high air velocity, low ambient temperature or direction of air flow, whereby more heat is withdrawn from a person's skin than is normally dissipated

draught diverter a device, without moving parts, fitted in the flue of an appliance for isolating the combustion system from the effects of pressure changes in the secondary flue

draught, forced combustion air supplied under pressure to the fuel burning equipment

draught, natural difference between atmospheric pressure and some lower pressure existing in the furnace or gas passages of heat generating unit, chimney effect

dress guard a grid fitted to a space heater to lessen the chance of fire or injury

duct a passageway made of sheet metal or other suitable material, not necessarily leaktight, used for conveying air or other gas at low pressures

ducted air heater a heater with a fan, made for the delivery of warm air by ducts

effect, chimney tendency of air or gas in a duct or other vertical passage to rise when heated due to its lower density compared with that of the surrounding air or gas; in buildings, tendency toward displacement (caused by the difference in temperature) of internal heated air by unheated outside air due to the difference in density of outside and inside air

electronic flame safeguard a flame safeguard utilising electronic components to perform its function. See also flame safeguard

evaporation change of state from liquid to vapour

excess air air in excess of that required for complete combustion which is mixed unchanged with the combustion products, in the combustion chamber

expansion, coefficient of the change in length per unit length or the change in volume per unit volume, per degree change in temperature

fan an air-moving device comprising a wheel or blade, and housing or orifice plate

fan, centrifugal a fan rotor or wheel within a scroll-type housing and including driving mechanism supports for either belt drive or direct connection

fan propeller a propeller or disc-type wheel within a mounting ring or plate and including driving mechanism supports for either belt drive or direct connection

fan shroud a protective housing which surrounds the fan and which may also direct the flow of air

flame abnormality a flame condition which results in lifting, floating, lighting back, appreciable yellow tipping, carbon deposition or objectionable odour

flame blow-off separation of a flame from a burner port, resulting in extinction

flame detector a device that is sensitive to flame properties and initiates a signal when flame is detected

flame establishment period the period which begins when the fuel valve is energised and ends when the flame supervision system is first required to supervise that flame

flame failure response time the time taken for the flame safeguard to detect loss of flame and de-energise the shut-off valve

flame lift separation of a flame from a burner port, whilst continuing to burn with its base some distance from the port

flame proving period the supervised period following the flame establishment period and before any further operation other than shut-down is permitted

flame safeguard a safety device which automatically cuts off the gas supply if the actuating flame is extinguished

flame safeguard system a system consisting of a flame detector(s) plus associated circuitry, integral components, valves and interlocks, the function of which is to shut off the fuel supply to the burner(s) in the event of ignition failure or flame failure

flame trap (flame arrestor) a device designed to prevent the flashing-back of flame through it under all conditions

flash point temperature of combustible material, such as oil, at which there is a sufficient vaporisation to support combustion of the material

flash tube a device for igniting a gas burner in which a flame is made to travel to the burner ports through a tube in which a flammable mixture of gas and air has been induced

flue the passage through which flue gases are conveyed from an appliance to a discharge point, excluding draught diverter, barometric device, fan or similar part

flue brick a hollow brick designed specially for use as a flue

flue connection a device incorporated in an appliance for the connection of a flue or draught diverter, barometric device, fan or similar part

   Note: Where the draught diverter, barometric device, fan or similar component is an integral part of the appliance, the discharge point of the integral part shall be deemed to be the flue connection

flue connection elbow a right-angle piece between the flue connection and the flue

flue cowl a fitting placed at the flue terminal, designed to prevent the entry of rain or the disturbing effect of wind while not interfering with the discharge of flue gases

flue gases combustion products plus all diluents and contaminants intermixed with the combustion products. These include, where applicable, excess air, dilution air, process air and waste products from the process

flue losses the sensible heat and latent heat above room temperature of the flue gases leaving the appliance

flue system the passage through which flue products are conveyed from an appliance to a discharge point including draught diverter, barometric device, fan or similar part

flue terminal the point at which flue products discharge from a flue. It is the point where a flue system discharges into a flue cowl (if fitted)

flued appliance an appliance designed to discharge its flue products through a flue connection

flueless appliance an appliance designed not to discharge its flue products through a flue connection

forced convection oven an oven in which the internal atmosphere is circulated by a fan

fryer an appliance for cooking food in hot oil

gas a combustible fuel gas which may be natural gas (NG): a hydrocarbon gas consisting mainly of methane; simulated natural gas (SNG): a gas comprising a mixture of LPG and air, in the approximate proportions of 55% gas and 45% air for commercial propane; town gas (TG): a gas manufactured from coal or petroleum feedstocks; tempered liquefied petroleum gas (TLP): a gas comprising a mixture of LPG and air, in the approximate proportions of 27% gas and 73% air for commercial propane; liquefied petroleum gas (LPG): a gas composed predominantly of any of the following hydrocarbons, or any combination of them in the vapour phase—propane, propene (propylene), butane, butene (butylene)

gas consumption the number of megajoules per hour introduced into an appliance under specified conditions

gasfitting work refer to local authority requirements and regulations

gas, high pressure gas supplied at pressures of over 200 kPa up to 1050 kPa

gas, low pressure gas supplied at pressures 7 kPa or less

gas, medium pressure gas supplied at pressures of over 7 kPa up to 200 kPa

gravity, specific density compared with density of standard material, it usually refers to water or to air

griddle a solid plate heated from below which is used to cook food directly on its surface. Also known as a dry-fry plate, hamburger plate or grill plate

grill dish or grill tray a dish or tray which catches any drips from the grill operation and contains the grill grid

grille a louvred or perforated covering for an air passage opening which can be located in the sidewall, ceiling or floor

grill fret externally heated radiant element of a griller

grill grid the wire work grid or perforated plate on which is placed the food to be cooked by a griller

griller an appliance for cooking food by radiant heat

gross oven volume the total volume of the oven between side walls, top and floor, inner door face and oven back panel

head, static the static pressure of a fluid expressed in terms of height of a column of the fluid, or of some manometric fluid, which it would support

heat form of energy that is transferred by virtue of a temperature difference

heat exchanger a device specifically designed to transfer heat between two physically separated fluids

heating value, gross (at constant pressure) the number of megajoules liberated when one cubic metre of dry gas, at standard conditions of temperature (15°C) and pressure (101.325 kPa), is completely burnt in air with all the water formed by the combustion process in the liquid state and with the products of combustion at standard conditions

heat, latent, of condensation difference is specific enthalpy of a condensible fluid between its dry saturated vapour state and its saturated liquid state at the same pressure

heat, latent, of condensation or evaporation (specific) thermodynamically, difference in the specific enthalpies of a pure condensible fluid between its dry saturated vapour state and its saturated (not subcooled) liquid state at the same pressure

heat, sensible heat which is associated with a change in temperature; specific heat exchange of temperature; in contrast to a heat interchange in which a change of state (latent heat) occurs

hose flexible tube or pipe of multiple construction

hose assembly flexible hose including end fittings connecting the gas supply to the appliance inlet

hotplate a hotplate contains one or more boiling burners which heat from underneath, by direct flame contact, bases of vessels placed over them. A griller may form part of the hotplate

humidifier a device to add moisture to air

humidify to add water vapour to the atmosphere, to add water vapour or moisture to any material

humidistat a regulatory device, actuated by changes in humidity, used for automatic control of relative humidity

humidity water vapour within a given space

humidity, absolute the weight of water vapour per unit volume

humidity, percentage the ratio of the specific humidity of humid air to that of saturated air at the same temperature and pressure, usually expressed as a percentage (degree of saturation; saturation ratio)

indirect fired oven an oven in which the products of combustion do not flow through the oven compartment

induced draught burner a burner where all or part of the air for combustion is introduced by providing suction in the combustion chamber by mechanical means

induced draught canopy a canopy in which the draught is induced by a fan

injector a device which causes air to mix with a stream of gas. In the case of an aerated burner it incorporates an orifice discharging gas into the mixing throat or tube

infiltration air flowing inward as through a wall, leak, etc.

instantaneous water heater a water heater in which gas, except for a pilot flame, burns only when water is being discharged

insulation joints joints and fittings designed to prevent the flow of electricity across the joint

insulation, sound acoustical treatment of fan housing, supply ducts, space enclosures and other parts of system and equipment to isolate vibration or to reduce noise transmission

insulation, thermal a material having a relatively high resistance to heat flow and used principally to retard heat flow

interlock a device or function which ensures that the operation of items of equipment is dependent upon the fulfilment of predetermined conditions by other items of equipment

intermittent pilot a pilot which is automatically ignited each time the burner is started, and which is automatically extinguished at the end of the main flame establishment period

interrupter screw a type of primary aeration control in an atmospheric burner, consisting of a screw which can be adjusted into the throat of the burner

jointing compound a manufactured liquid, paste-like or semi-solid material used to seal joints

joint, welded a gas-tight joint obtained by joining of metal parts in the plastic or molten state

lateral a horizontal section of piping from or to a riser

leisure-type appliance an appliance operating on LPG normally used for cooking, lighting or heating in the outdoor environment and specifically intended not to be used inside a building, caravan, motor vehicle, marine craft, aircraft or within any confined and unventilated areas or where prohibited from use by local authorities

light back transfer of the flame from burner port(s) into the body of the burner or back to the injector

limit device a device which is actuated by the approach to a hazardous situation in an appliance due to abnormal conditions and, when actuated, causes the gas supply to all burners to shut off with lockout

lining blocking of a burner port or air opening by dust or fluff

lockout a safety shut-down condition of the control system which requires a manual reset in order to restart

lock-up pressure the outlet pressure of a pressure regulator which causes the regulator to shut off completely

louvre an assembly of sloping vanes intended to permit air to pass through and to inhibit transfer of water droplets

low-gas pressure detector a pressure sensing device which is actuated when the gas pressure falls below a pre-set value

LPG see gas

mains pressure water heater a water heater directly connected to the water mains and subjected to mains pressure

manometer an instrument for measuring pressures, essentially a U-tube partially filled with a liquid, usually water, mercury or a light oil, so constructed that the amount of displacement of the liquid indicates the pressure being exerted on the instrument

manual ignition the lighting of gas at a burner by a manual operation whenever gas flows from the burner

master meter a meter connected directly to the service pipe and supplying gas to subsidiary meters

micron a unit of length, the thousandth part of one millimetre or a millionth of a metre

mixing tube that part of an atmospheric burner in which the air and gas are mixed

mobile appliance an appliance with one or more wheels designed to be easily moved from place to place by one person

natural draught the flow produced by the tendency of warmed gases to rise natural draught burner *see* atmospheric burner

natural draught flue a flue which is operated only by the buoyancy effect of the hot gases in it

natural gas *see* gas

net oven volume the usable volume of an oven, calculated from the *height* (measured from the lowest portion of the oven on which a food container can be placed, usually the oven floor, to the lowest portion of any thermostat or other projection which can interfere with the insertion of food, or to the top of the door opening, whichever is lowest), *width* (measured between the shelf supports) and *depth* (measured from the leading edge of the grid to the unrestricted depth to which the tray may be introduced)

nominal gas consumption the appliance manufacturer's gas consumption, in megajoules per hour, stated in specifications, data plates, instructions and general communications

nominal outlet pressure the outlet pressure or outlet pressure range of a regulator, as specified by the manufacturer

nominal test point pressure the gas pressure specified by the manufacturer, at which the appliance is designed to operate, and measured at a test point

non-return valve a device designed to operate automatically to prevent reversal of flow in a pipe

open top boiling table a boiling table which has trivets

operating authority the organisation responsible for the design, construction, testing, inspection, operation and maintenance of facilities within the scope of the code

over-pressure cut-off device a mechanical device incorporated in a gas pipework system to shut off the supply of gas when the pressure at the sensing point rises to a predetermined figure

over-pressure protection a device preventing the pressure in piping or in appliances from exceeding a predetermined value

over-temperature cut-off device a manual reset or non-resetting device which functions to shut off the gas supply to a burner or burners to prevent temperature exceeding a predetermined level

over-temperature limiting device *see* relief device

oxygen depletion pilot (ODP) a pilot and safety device which causes the gas supply to be shut off when the oxygen content of the ambient air is depleted to a specified concentration

part automatic burner a burner system which includes any self-acting operation

permanent pilot a pilot which is permanently alight while the appliance is in service and which is controlled independently of the main burner

piezo electric ignition a form of ignition with which a high voltage spark is produced as the result of pressure or impact upon specially treated ceramic material

pigtail a short length of small bore pipe, usually annealed copper, used for the high pressure connection between an LPG cylinder(s) and the regulator, and looped or formed in a shape to minimise work hardening

pilot a permanently located burner independent of the main burner, small in relation to it, and arranged to provide ignition for the main burner

POL connection (Prest-O-Lite) the common name for a standard union with left-hand thread, used for connecting pigtails to LPG cylinder valves

portable appliance an appliance designed to be carried by the user from place to place, as required

power flued a flue system in which products of combustion are removed from the appliance by a fan in the flue

pressure, absolute pressure referred to a perfect vacuum. It is the sum of gauge pressure and atmospheric pressure

pressure, saturated the saturation pressure for a pure substance for any given temperature is that pressure at which vapour and liquid, or vapour and solid, can coexist in stable equilibrium

pressure, static 1 the pressure with respect to stationary surface tangent to the mass flow velocity vector; 2. the pressure with respect to a surface at rest in relation to the surrounding fluid

pressure drop static pressure loss in fluid pressure, as from one end of pipe to the other, due to friction, etc.

primary aeration control an adjustable device for controlling primary air or an atmospheric burner

primary air that portion of the total air required for combustion which is mixed with the gas prior to burning on the burner ports

primary air ports the openings in an aerated burner for primary air

primary flue that portion of the flue system conveying flue products from an appliance to the draught diverter, barometric device, fan or similar part when fitted

programmed ignition a multistage system of automatic ignition where each stage takes place automatically in a predetermined sequence

programming flame safeguard a flame safeguard which automatically sequences at least two burner functions such as ignition spark, gas valve, etc.

protected pilot a pilot which is fitted with a flame safeguard

psychrometer instrument for measuring relative humidities by means of wet-bulb and dry-bulb temperatures

purge to replace air or other gases which may become a hazard from pipework, fittings or appliances

quick-connect device a push-in or bayonet type end fitting which connects a flexible hose to the gas supply. It contains safety features which prevent gas flowing unless the connection is securely made

radiant convection heater a space heater for which the heat output consists of substantial proportions of warm air and radiation

radiant heater a space heater which has an output mainly of radiation

range a combination of individual commercial catering units

rated working pressure the maximum allowable inlet pressure specified by the manufacturer

recessed wall heater a flued convection heater designed for installation partly or wholly inside the room wall

refrigerant the fluid used for heat transfer in a refrigerating system, which absorbs heat at a low temperature and a low pressure of the fluid and rejects heat at a higher temperature and a higher pressure of the fluid, usually involving changes of state of the fluid

register a combination grille and damper assembly covering an air opening

regulator a device which automatically regulates the outlet pressure or the outlet volume of gas passing through it to a predetermined value. Regulators may be adjustable (usually spring loaded) or non-adjustable (usually weight loaded)

regulator capacity the maximum flow rate of the regulator as specified by the manufacturer

regulator, draught a device sometimes installed in the breeching between a fired appliance and the chimney; it is intended to control chimney draught effect on inducing gas flow through the appliance

regulator range the manufacturer's specified range of flow rate for which the regulator is suitable

relative density the density of a known volume of dry gas relative to that of the same volume of dry air under the same conditions of temperature and pressure

relay valve a diaphragm-operated valve with a controlled weep flow rate which governs the action of the valve

relief device a safety device designed to forestall the development of a dangerous condition by preventing pressure, temperature or vacuum build-up

room sealed appliance an appliance designed such that air for combustion does not enter from, or combustion products enter into, the room in which the appliance is located

rotisserie a rotating spit, manual or automatic, enclosed or otherwise, heated by radiation and/or convection

safety shut-off system an arrangement of valves and associated control systems which shuts off the supply of gas when required, by a device which senses an unsafe condition

safety shut-off valve an automatic shut-off valve used to shut off gas supply to an appliance when a signal is generated indicating that a dangerous condition has developed

salamander a griller provided with an adjustable grid

sealing plate a plate used to support and seal the transition piece to the flue. It prevents the entry of excess air into the flue

secondary air air required for completion of combustion, admitted to the combustion zone after combustion with primary air has commenced

secondary flue that portion of the flue system conveying flue products from the draught diverter, barometric device, fan or similar part to the flue terminal

semi-automatic ignition the lighting of gas at a burner using a combination of automatic and manual operations

service the pipe which runs between a main and a consumer's meter

service riser that section of the service pipe connecting from below ground to the service regulator and/or consumer's meter

specific gravity *see* relative density

spillage a condition in which the combustion products are not cleared by the flue, but escape from the appliance or the draught diverter relief openings

statutory authority the state or Commonwealth body empowered by an Act of parliament to exercise jurisdiction over facilities within the scope of the code

stock pot a vessel designed for the bulk cooking of liquid foodstuffs such as soup, stock and porridge

stoichiometric combustion combustion of a fuel–air mixture in exact weight proportion so that the theoretical products of combustion are only carbon dioxide, water vapour and the nitrogen in the combustion air and no excess oxygen or incomplete combustion products

storage water heater a water heater in which a hot water storage vessel is built into the appliance as an integral part

subsidiary meter a meter for measuring gas usage in separate parts of a premises or in separate appliances, and which has already passed through a master meter

surface combustion burner a burner where combustion takes place on the surface of a perforated or porous material or gauze

temperature the thermal state of matter with reference to its tendency to communicate heat to matter in contact with it

temperature, absolute temperature expressed in kelvins

temperature, absolute zero the zero point on the Kelvin temperature scale, $-273.15°$ C $(-459.69°$ F)

temperature, dewpoint see dewpoint temperature

temperature, dry-bulb the temperature of a gas or mixture of gases indicated by an accurate thermometer after correction for radiation

temperature, wet-bulb thermodynamic wet-bulb temperature is the temperature at which liquid or solid water, by evaporating into air, can bring the air to saturation adiabatically at the same temperature. Wet-bulb temperature (without qualification) is the temperature indicated by a wet-bulb psychrometer constructed and used according to specifications

temperature, wet-bulb depression difference between dry-bulb and wet-bulb temperatures

tempered liquefied petroleum gas *see* gas

terminal *see* flue terminal

temperature limit device a device which automatically interrupts the heat flow when the temperature at the control point reaches a predetermined limit

thermocouple a device that utilises the fact that an electromotive force is generated whenever two junctions of two dissimilar metals in an electric circuit are at different temperature levels

thermometer an instrument for measuring temperature

thermostat an automatic control device actuated by temperature and designed to be responsive to temperature

thermostat a device which automatically maintains a predetermined temperature in an appliance or component

thermostat bypass an integral part of a thermostat control which enables a preset volume of gas to bypass the thermostat valve

throw the horizontal or vertical axial distance an air stream travels after leaving an air outlet before the maximum stream velocity is reduced to a specified terminal level

trivet grid located over the hotplate burners to support vessel being heated

under-pressure cut-off device a mechanical device incorporated in a gas pipework system to shut off the supply of gas when the pressure at the sensing point falls to a predetermined figure

unvented water heater a water heater in which no provision is made for a vent permanently open to atmosphere

valve a device for the purpose of controlling or shutting off flow

valve, pressure reducing a valve which maintains a uniform pressure on its outlet side irrespective of how the pressure on its inlet side may vary above the pressure to be maintained

valve, pressure relief a valve held closed by a spring or other means and designed to automatically relieve pressure in excess of its setting; also called a safety valve

valve, solenoid a valve which is closed by gravity, pressure or spring action and opened by the movement of a plunger due to the magnetic action of an electrically energised coil, or vice versa

valve, stop a shut-off valve, other than a valve for controlling the flow

valve train a number of valves, regulators, pipe pieces, unions, etc. which form an integrated system for flow and/or pressure control and safe operation of a burner

vented water heater a water heater in which provision is made for a vent permanently open to the atmosphere

vent line a pipe or tube which conveys gas to a safe place outside the building from a gas pressure regulator relief valve, or a double block and vent safety shut-off system

ventilation the process of supplying or removing air by natural or mechanical means to or from any space. Such air may or may not have been conditioned

volume, specific the volume of a substance per unit mass; the reciprocal of density

water heater an appliance for the supply of water at a temperature below 99°C

water-operated gas valve a gas valve actuated by a flow or change of pressure in a water system of an appliance

water pressure reducing valve a valve which automatically reduces inlet water pressure to a specified value at its outlet

water pressure relief valve a pressure-actuated valve which automatically discharges fluid at a specified set pressure. It is fitted to heaters to prevent the pressure in the container from exceeding the maximum working pressure during normal use and from exceeding a predetermined margin over the maximum working pressure when temperature controls fail to limit pressure

Index

Exercises

chapter one

1 How many cubic metres (m³) of natural gas would be required to supply 133 MJ of energy?

2 State the volume of air that would be required to burn completely 1.5 m³ of propane.

3 Why was towns gas toxic?

4 The natural gas distribution system does not depend on gas holders for the storage of gas. State the reason for this.

5 Name the gases used as a feedstock to produce TLP and SNG.

6   What Australian Standards are used for gas installations?

_____

_____

_____

7   What is the statutory authority in your area for LPG tank installation?

_____

_____

_____

8   What are the gas pressures in low, medium and high pressure supply mains?

_____

_____

_____

_____

chapter two

1 (a) Describe the difference between static pressure and working pressure.

_____

_____

_____

_____

_____

(b) Discuss where you would connect the manometer to check the working pressure of a cooker.

_____

_____

_____

(c) Could you check the static pressure with the manometer connected in the position you indicated when answering question 1(b)?

_____

_____

_____

2 When testing a domestic service for soundness, that has appliances connected, should the appliances be isolated? If so, why?

_____

_____

_____

_____

_____

3 Convert the following pressures, giving all answers in kPa.

(a) 10 psi   _____

(b) 500 mb   _____

4   What effect, if any, would an increase in altitude have on gauge pressure if the gas being used was natural gas?

_____

_____

_____

_____

5   Where do we connect the manometer to set the regulator appliance pressure?

_____

_____

_____

6   Where do we connect the gauge to get the pressure in the main?

_____

_____

_____

chapter three

1   Name the products of complete combustion.

2   State the causes of incomplete combustion.

3   What is stoichiometric combustion?

4   Draw an aerated burner, name the parts and describe its operation.

5   Where does an aerated burner get its air for combustion?

6 (a) Why is flame retention necessary?

_____

_____

_____

(b) Describe how a flame retention port works.

_____

_____

_____

7 Describe the flame characteristics of an aerated burner under the following conditions:

(a) lack of primary air

_____

_____

_____

(b) too much primary air

_____

_____

_____

(c) correct air–gas mix.

_____

_____

_____

8 What is excess oxygen and what is it used for?

_____

_____

_____

chapter four

1   State three factors affecting the inlet pressure to a service regulator.

2   With the aid of a drawing, describe the operation of a service regulator.

3   Describe the operation of the pressure relief valve section of a medium to low pressure service regulator.

4   Complete Table 4.1 with regard to a medium to low pressure service regulator.

Table 4.1		
Fault	**Causes**	**Remedy**
Gas issuing from vent	1. 2.	
High lock-up pressure	1. 2.	
Regulator chatter	1. 2.	

5   State two advantages of a spring-loaded regulator.

6   Why is it important to use the correct regulator spring?

7   State three factors affecting the inlet pressure to an appliance.

8   With the aid of a drawing describe the operation of an appliance regulator.

9   Explain how you would adjust the working pressure of an appliance regulator.

10 What is two stage regulation and what advantage does it offer?

11 When would an installation require over-pressure protection?

chapter five

1   List two functions of a thermostat.

2   Explain the operation of a snap action rod and tube thermostat.

3   Name an appliance that may use a snap action rod and tube thermostat.

4   Explain the operation of a liquid expansion thermostat.

5   State the main advantage that the liquid expansion thermostat has over a rod and tube thermostat.

6   What appliance uses a graduating liquid expansion thermostat?

7 What appliance may use a snap acting liquid expansion thermostat?

8 Explain the operation of a vapour pressure thermostat.

9 What appliance uses a vapour pressure thermostat?

10 What function does the burner play when used on a refrigerator?

11 Describe the operating procedure of the bimetal thermostat when controlling the operation of the fan and high limit switch in a warm air ducted central heating unit.

12 Why are heat anticipators used on room thermostats?

chapter six

1 Where were bimetallic strips used?

2 What is a thermocouple?

3 What is a thermopile?

4 What device do we use to check the operation of thermoelectric flame failure devices?

5 What would be the effect on the operation of a thermoelectric flame failure device if the thermocouple were exchanged for one with a larger mass?

6   Draw a simple line diagram, name the parts and explain the principle of operation of a thermoelectric flame failure device.

7   Briefly describe how flame rectification works.

8   What device shuts off the gas flow if the flame rectification circuit is incomplete?

9   If the flame rod was making metal-to-metal contact with the burner on a flame rectification circuit, what effect, if any, would this have on the gas supply?

10 Draw a simple flame rectification circuit.

11 Describe the operation of an ultraviolet photoelectric cell.

_____

_____

_____

_____

_____

_____

12 State the advantages of the ultraviolet photoelectric cell when compared with the thermoelectric flame failure device.

_____

_____

_____

_____

_____

13 What two sensing devices are used in high temperature cut-outs?

_____

_____

_____

14 How does a high temperature cut-out shut off the gas supply?

_____

_____

_____

_____

_____

15 Why would we want the gas supply to shut off if gas pressure was too low or too high?

_____

_____

_____

_____

_____

16 How does a pressure cut-off device work?

_____

_____

_____

_____

_____

17 State the function of an oxygen depletion device.

_____

_____

_____

_____

_____

chapter seven

1   Describe the operation of a flash tube ignition system.

2   State two reasons why flash tube ignition is not effective with natural gas appliances.

3   Describe the operation of a piezo ignition system.

4   What would you check if a piezo ignition system failed to produce a spark?

5   Describe the operation of an electronic ignition system.

6 Name two advantages which a mains electricity electronic igniter has when compared with a piezo igniter.

7 Describe the operation of the additional features of a re-igniter.

8 If ignition does not occur in an oven, how does the re-ignition device react?

9 Why is it important to maintain the correct spark gap?

chapter eight

1   Describe the operation of a solenoid valve.

2   Why are DC solenoids less susceptible to solenoid hum?

3   (a) If the adjustable bypass on a relay valve were set too low, what effect, if any, would this have on the operation of the relay
         valve?

    (b) If a thermostat fitted into the weep tube of a relay valve failed to seal when the set temperature was reached, what effect
         would this have on the supply of gas to the burner?

4   Name three control devices commonly found in combination controls:

1   _____

    _____

2   _____

    _____

3   _____

    _____

5   State the function of each of the control devices in question 4:

1   _____

    _____

2   _____

    _____

3   _____

    _____

chapter nine

1   State two functions of a flue.

2   List four materials that can be used for a natural draught flue, including their limitations.

3   State four causes of down draught in a natural draught flue.

4   Draw a natural draught flue system. Name the parts and describe their function.

5   Describe the operation of a balanced flued appliance.

6   State three factors of appliance performance that would be affected by poor ventilation.

7   What is the main difference between a natural draught and a power flue?

8   What is the purpose of interlocking the gas supply with a power flue fan?

9   Complete the following table by indicating the minimum distance for a natural draught balanced flue terminal from the following (refer to AS 5601):

Appliance nominal thermal input/gas consumption	Minimum distance		
	Window	Gas meter	Electricity meter
Up to 150 MJ			
150 MJ to 200 MJ			

chapter ten

Using AS 5601, answer the following questions:

1   List four permitted materials and joining methods for consumer piping.

2   What is the minimum depth of consumer piping in a private property?

3   Briefly describe how you would purge a branch line with an appliance connected that was fitted with a thermo-electric flame failure device.

4   Briefly describe how you would test a branch line with an appliance connected that does not have a thermoelectric flame failure device.

5   From the information given below, calculate the main line and all branch line diameters for each pressure drop indicated on the table.

Type of gas: Natural. *HV*: 38   *RD*: 0.6   Material: Copper.

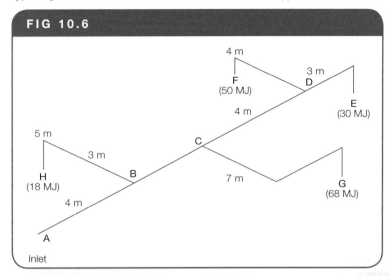

FIG 10.6

Pipe sizing chart for Figure 10.6		Supply pressure around 1.1 kPa	Supply pressure around 1.25 kPa	Supply pressure 1.5–2.5 kPa
		Press. drop: 0.075	Press. drop: 0.12	Press. drop: 0.25
		Table no.	Table no.	Table no.
Pipe section	Gas rate	Pipe size	Pipe size	Pipe size

6   From the information given below, calculate the main line and all branch line diameters for the LPG installation in Figure 10.7.

Type of gas: Propane. *HV*: 96   *RD*: 1.5   Material: Copper.

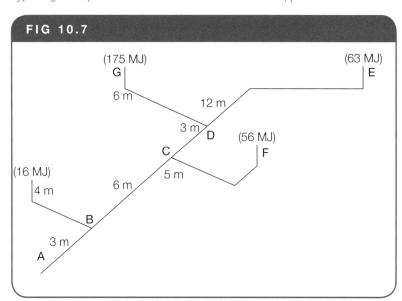

FIG 10.7

(175 MJ)
G

(63 MJ)
E

6 m   12 m

3 m D

C   (56 MJ)
F

(16 MJ)

4 m   6 m   5 m

B

3 m

A

Pipe sizing chart for Figure 10.7		Press. drop: 0.25
		Table no.
Pipe section	Gas rate	Pipe size

7   From the information given in Figure 10.8, calculate the main line and all branch line diameters for the first stage of the LPG installation.

Type of gas: Propane. *HV*: 96   *RD*: 1.5   Material: Copper. First-stage regulator: 70 kPa; second-stage regulator: 3 kPa.

Use charts provided in the AS 5601 code. Refer to Figure 10.8.

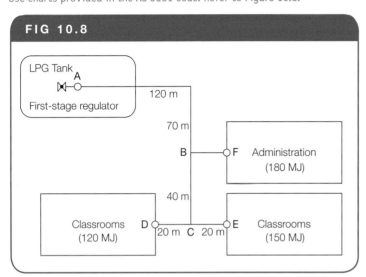

FIG 10.8

LPG Tank
A
First-stage regulator
120 m

70 m

B   F   Administration
(180 MJ)

40 m

Classrooms   D
(120 MJ)   20 m C 20 m   E   Classrooms
(150 MJ)

Index length = _____ metres

Pipe sizing chart for Figure 10.8		
Section	**MJ loading (MJ/Hr)**	**Pipe size (mm)**
A-B		
B-C		
B-F		
C-D		
C-E		

chapter eleven

1 With the aid of a sketch describe the operation of a fan-forced oven.

2 (a) What is meant by the term 'oven vitiation'?

_____

_____

(b) State three causes of oven vitiation.

1 _____

2 _____

3 _____

3 Give the reasons why it is essential that an instantaneous hot water service must light up in two stages.

_____

_____

_____

_____

_____

4   List the commissioning procedure for an upright stove.

Installation check

1 _____

2 _____

Light up and check

1 _____

2 _____

3 _____

4 _____

5 _____

6 _____

7 _____

8 _____

9 _____

10 _____

Instructions for customer

1 _____

2 _____

3 _____

5   Where and when would you check the water temperature to determine whether or not the thermostat of a storage hot water service needed recalibrating?

_____

_____

_____

6   (a) What are the advantages of radiant heat when used to heat a room?

_____

_____

_____

_____

(b) What are the advantages of convection heat when used to heat a room?

_____

_____

_____

_____

7   Why is an oxygen depletion device fitted to flueless LPG room heaters?

_____

_____

_____

_____

8   How is the temperature maintained on a continuous flow water heater if the water flow rate changes?

_____

_____

_____

_____

9   If after checking the appliance performance against the rating on the data plate it was found to be outside the 10% range what would you do?

_____

_____

_____

_____

10 Describe what maintenance procedures you would perform when servicing a natural gas storage hot water service.

_____

_____

_____

_____

_____

11 List and describe the three methods of heat transfer.

1 _____

_____

2 _____

_____

3 _____

_____

12 Why is it important to instruct the customer on the operation of the appliance and leave the instructions with them?

_____

_____

_____

_____

_____

1   State the degree of accuracy that a meter must be within to be approved and stamped.

2   Describe how a dry diaphragm meter operates.

3   Where would a rotary meter be used?

4   Describe how a rotary meter operates.

5 Why are bonding straps used when disconnecting meters?

6 On a domestic property where is the gas meter usually located?

7 On a domestic property why are the gas risers usually metallic?

8 How would you know if the meter was of adequate capacity for the installation?

9 Sketch the connection of a meter within your area, including inlet and outlet piping.

10 List six prohibited locations for a gas meter (refer to AS 5601).

1 _____

2 _____

3 _____

4 _____

5 _____

6 _____

11 When checking the gas rate of an upright cooker the test dial of the meter recorded 20 litres in 43 seconds. How much gas is the appliance consuming and is it within 10% of the appliance's nominal thermal input of 64 MJ/hr?

$$GR = \frac{V \times H \times HV}{T} \text{ (explain the formula)}$$

Where:

GR = _____

V = _____

H = _____

HV = _____

T = _____

$$GR = \frac{HV \times H \times V}{T}$$

$$= \frac{\quad \times \quad \times \quad}{\quad}$$

Answer: _____

Is this within 10 %? _____

12 When checking the gas rate of a space heater the following results were recorded. The test dial of the meter recorded 0.01 cubic metres in 1 minute and 10 seconds. How much gas is the appliance consuming and is it within 10% of the appliances nominal thermal input of 18 MJ/hr?

$$GR = \frac{V \times H \times HV}{T}$$

$$= \frac{\quad \times \quad \times \quad}{\quad}$$

Answer: _____

Is this within 10 %? _____

13 When checking the gas rate of a continuous flow heater the following meter readings were recorded.

2	1	6	8	1	4	5

Before test

2	1	6	8	1	7	5

After test

If the test took 23 seconds how much gas was the appliance consuming and is it within 10% of the appliance's nominal thermal input of 180 MJ/hr?

$$GR = \frac{HV \times H \times V}{T}$$

$$= \frac{\quad \times \quad \times \quad}{\quad}$$

Answer: _____

Is this within 10 %? _____

chapter thirteen

1 Name two factors that affect the vaporisation rate of propane.

2 When no gas is being used, what prevents the gas from vaporising in a cylinder?

3 One volume of liquid propane will vaporise to form approximately how many volumes of gas?

4 Cylinders under heavy load sometimes have frost on the outside. Why?

5 List the safety factors to be observed when transporting propane cylinders.

6   Give two reasons why a pressure relief valve on an LPG cylinder may operate.

7   Why are LPG cylinders filled only to the 83% level?

8   (a) Why is it standard practice to run a medium pressure line between a bulk tank and the second-stage regulator? (b) Why must the second-stage regulator be fixed outside the building?

1   State the precautions you would take to eliminate ignition sources when entering a gas-filled building.

2   What are the physical signs shown by a person suffering from carbon monoxide poisoning?

3   What action should be taken if a 45 kg cylinder's pressure relief valve was blowing? Assume the gas to be alight and burning away from the building and no flame contact with the cylinder.

4   What are the hazards of escaping unburnt natural gas?